American Institute of Chemical Engineers

Liaison Functions Conference

Presentations at the
2008 AIChE Spring National Meeting

April 6-10, 2008
New Orleans, Louisiana, USA

Printed from e-media with permission by:

Curran Associates, Inc.
57 Morehouse Lane
Red Hook, NY 12571
www.proceedings.com

ISBN: 978-1-60560-222-6

Some format issues inherent in the e-media version may also appear in this print version.

ISBN: 978-1-60560-222-6

American Institute of Chemical Engineers
Proceedings
Three Park Avenue
New York, NY 10016-5991
Phone: 212-591-8100

www.aiche.org

American Institute of Chemical Engineers

Liaison Functions Conference
2008

TABLE OF CONTENTS

22a Progress and New Developments in CO_2 Capture and Storage

Sean Plasynski, Strategic Center For Coal R&D, National Energy Technology Laboratory, 626 Cochrans Mill Road, PO Box 10940, Pittsburgh, PA 15236, Derek M. Vikara, Science Applications International Corporation, National Energy Technology Laboratory, PO BOX 10940, 626 Cochrans Mill Road, Pittsburgh, PA 15236, and **Rameshwar Srivastava**, Science Applications International Corporation, National Energy Technology Laboratory, 626 Cochrans Mill Road, P.O. Box 10940, Pittsburgh, PA 15236.

Carbon capture and storage (CCS) is an approach aimed at mitigating the effects of global climate change by capturing carbon dioxide (CO_2) from large point sources and subsequently storing it instead of releasing it into the atmosphere. As the nation's energy demand continues to increase, so will the use of fossil fuels. Despite gains in overall energy efficiency and increased use of low- or no-carbon fuels, CCS technologies will likely be essential in the effort to stabilize atmospheric concentrations of greenhouse gases (GHG).

There has been a dramatic expansion of interest, both nationally and worldwide, in the CCS program and the potential to meet our energy and environmental challenges. The U.S. Department of Energy (DOE) established the Carbon Sequestration Program in 1997 upon recognizing the importance of CCS for global climate change mitigation. DOE's Carbon Sequestration Program is using applied research with field demonstration to assess the technical and economic viability of CCS to reduce GHG concentrations in the atmosphere. The objective of CO_2 capture research is to cost-effectively produce a CO_2-rich stream at pressure for the three possible capture pathways: post-combustion (recovery of CO_2 from flue gas), pre-combustion (recovery of CO_2 from fuel gas), and oxyfuel (combustion in pure O_2 to produce flue gas that is nearly pure CO_2). Advances in CO_2 capture and separation technology will be discussed. The DOE Program has a variety of projects testing geologic carbon sequestration focusing on safe and cost-effective storage of captured CO_2. Field projects are investigating sequestration in oil and gas reservoirs, saline formations, and deep coal seams. DOE's seven Regional Carbon Sequestration Partnerships are currently moving forward with priority sequestration technology field validation tests by implementing large-scale CO_2 injections coupled with monitoring efforts to ensure stable geologic and surface conditions and modeling to predict the fate and transport of injected CO_2 far into the future.

Polyimides and Their Derivatives for Gas Separation Applications

John R. Klaehn[1], Christopher J. Orme[1], Thomas A. Luther[1], Eric S. Peterson[1], Jagoda M. Urban-Klaehn[2]

[1]*Idaho National Laboratory, P.O. Box 1625, Idaho Falls, ID 83415-2208*
[2]*Idaho Accelerator Center, 1500 Alvin Ricken Dr., Pocatello, ID 83201*

Introduction

Fossil fuels will remain the core of energy production well into the 21st century. Availability of these fuels to provide clean, affordable energy is essential for the prosperity and security of the United States. Roughly one third of the United States' carbon emissions come from utilities, such as power plants. To stabilize and ultimately reduce greenhouse gas concentrations (especially carbon dioxide), DOE has established the Carbon Sequestration Program with the charge to develop methods for carbon emissions reductions. To be successful, carbon sequestration must meet the following requirements: 1) have effective separation and be cost-competitive, 2) have stable, long term storage, and 3) be environmentally benign.

Polymer membranes are an effective, economic, and flexible tool for many gas separations. For example, the commercial use of polymer membranes for air separation, the recovery of hydrogen from nitrogen, carbon monoxide, and methane mixtures, and the removal of carbon dioxide from natural gas have been widely reported. In each of these applications, high fluxes and excellent selectivities have relied on glassy polymer membranes. To date, this technology has focused on optimizing materials for near ambient conditions. Therefore, the development of high temperature polymeric membranes for CO_2 separations would represent a vital contribution for economic and environmental reasons.

A set of polyimide-based materials has been identified as an excellent membrane material. The VTEC polyimides (Kapton-like) possess thermal properties (up to 500°C) and these polyimides are robust and flexible even after multiple thermal cycles (up to 400°C). These polymers have proven to be very interesting for the carbon capture and sequestration effort. Another aspect of the high performance polymers is polybenzimidazole (PBI, Figure 1). It is another attractive polymer for challenging high temperature gas separation applications.[1,2] The INL has developed synthetic routes for a range of post-polymerization-modified PBI materials. Modification of the base PBI has been accomplished with polymer solubility in common organic solvents accomplished. This result makes possible application of membranes to complex shapes utilizing common membrane casting methods. It is important to note that the substituted PBI polymers were not advantageous for high temperature gas separation; however, these substituted polymers have been shown to be useful as blends with VTEC.

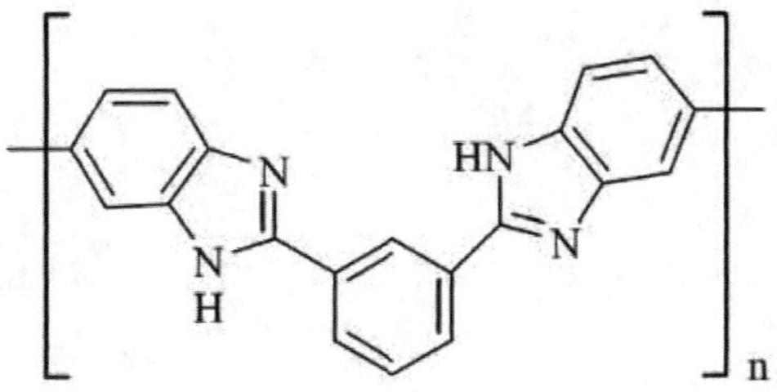

Figure 1. PBI (polybenzimidazole)

Experimental

Instrumentation. Proton and $^{13}C\{^1H\}$ NMR spectra were recorded on a Bruker DMX 300WB spectrometer operating at 300 MHz (^1H), 75 MHz (^{13}C). Thermal analyses were obtained using TA Instruments Model 2910 differential scanning calorimeter (DSC) and a Model 2950 thermogravimetric analyzer (TGA).

Pure Gas Testing/Pressure Rise Time-Lag method. The gas-testing results were obtained using the pressure-rise time-lag method.[3] Membranes were exposed to six different gases: H_2, N_2, O_2, Ar, CH_4 and CO_2. In a typical experiment, both sides of the membrane are evacuated to an equal vacuum. The test cell is then isolated and the pressure at zero time is used as the baseline. Next, the feed side is exposed to the test gas. Finally the pressure build up on the permeate side of the membrane is recorded as a function of time. The two gas transport properties that are determined directly from the pure gas test system are time lag and permeability. The permeability is the rate at which the gas permeates through the membrane after the gas has come to equilibrium in the polymer. The time lag is the time it takes the gas to permeate from the feed side of the membrane to the permeate side and is used to calculate the diffusivity.

The basic relationship of gas-transport properties, permeability (P), solubility (S), and diffusivity (D) in polymeric membranes is expressed in the following terms.

$$P = DS \quad (1)$$

Permeabilities for these experiments were calculated using the following equations where test system volume is V (cm^3), test system feed gas initial pressure is p_1 (cm Hg), test system temperature is T (°C), membrane thickness is l (cm), and membrane area exposed to the feed gas is A (cm^2). The slope was determined from a least squares fitted line of the time (sec) versus permeate gas pressure (cm Hg) steady-state data set (data during equilibrium flux).

$$P = slope \frac{V}{76} \frac{273}{(273+T)} \frac{1}{A} \frac{l}{p_1} \quad (2)$$

The time-lag is the intercept of the baseline and the least square fitted line for the data set. The determined value for the time lag is used to calculate the diffusivity D.

$$t = timelag = \frac{baseline - intercept}{slope} \quad D = \frac{l^2}{6t} \qquad (3)$$

The gas solubility is algebraically calculated from the measured quantities of permeability and diffusivity using eq 1.

Discussion

From previous work, INL has successfully synthesized, characterized and tested various N-substituted PBI polymers where Z = RMe_2SiCH_2-, R = alkyl, phenyl (see Scheme 1).[4] This post-polymerization polymer modification route provides a straightforward synthetic method that can be carried out at room temperature and give moderate yields. These N-substituted PBI polymers are soluble in common solvents, such as tetrahydrofuran (THF) and chloroform which provides materials that could be more amenable to processing. Based upon the NMR analysis in CDCl3, nearly all of the benzimidazole N-H positions (two positions per repeat unit) are substituted by the organosilane moieties. Interestingly, some of the modified polymers have similar thermal properties compared to the parent polymer, and the average molecular weights are higher for the substituted polybenzimidazoles than the parent PBI.

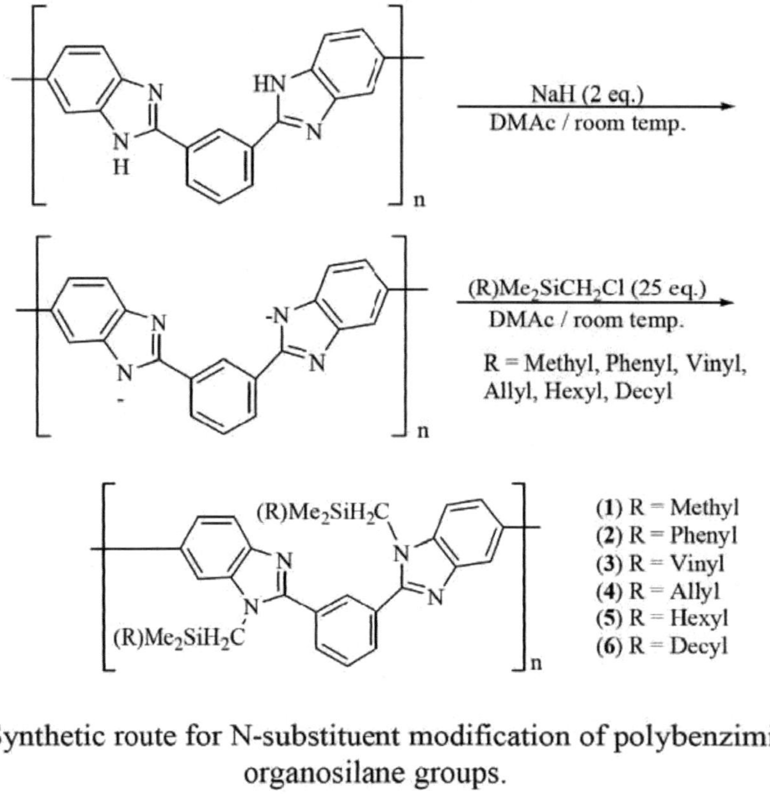

Scheme 1. Synthetic route for N-substituent modification of polybenzimidazole with organosilane groups.

There are few literature citations that describe N-substituted PBI materials that are soluble in common solvents.[5] Our purified modified polymers will dissolve in THF and chloroform. The yields among the substitutions were approximately 50%. Overall, our gas testing data for N-substituted PBI polymers show that the gas permeabilities were higher, but the gas separations (like H_2/CO_2) were not as good as the parent PBI material.

A promising family of polymers for gas separation is VTEC polyimides. We are using VTEC polymers along with N-substituted PBI and other polymer systems to develop membranes that can provide better gas separation. Interestingly, VTEC polymer films do not change after being heated in a convection oven at 250 °C for days. Conversely, many organic polymers will degrade (carbonize) at high temperatures when exposed to the atmosphere for extended periods. The gas transport data from the polymers is quite astonishing and competitive with the other high performance gas separation polymers. As we began to observe the transport properties of the polymers, it quickly became apparent that the polymer processing procedures for preparation of the membranes are critical. Thus, we developed a heat treatment regimen that gives reliable gas testing data and repeatable film casting techniques.

TGA data (Figure 2) were taken on the same VTEC PI 80-051 with and without ~10 wt% elemental bromine added to the polymer. Over the 12-hour period of the experiment, at a temperature of 300°C, (all INL polymers were heat cycled before the TGA evaluation), very little weight loss was shown. The parent VTEC PI 80-051 shows a 1.06% weight loss and the same material with 10-wt% bromine at 300°C over 12 hours demonstrates only 1.5% weight loss. The weight loss difference between the two films is minor, less than 0.5%; therefore, the loss of the "remaining" bromine is not an issue. The team suggests that the "remaining" bromine is tightly held by the polymer by either van der Waals interactions or covalent bonding. In addition, several other INL's experiments have shown no problems with the VTEC polymers operating at 300°C.

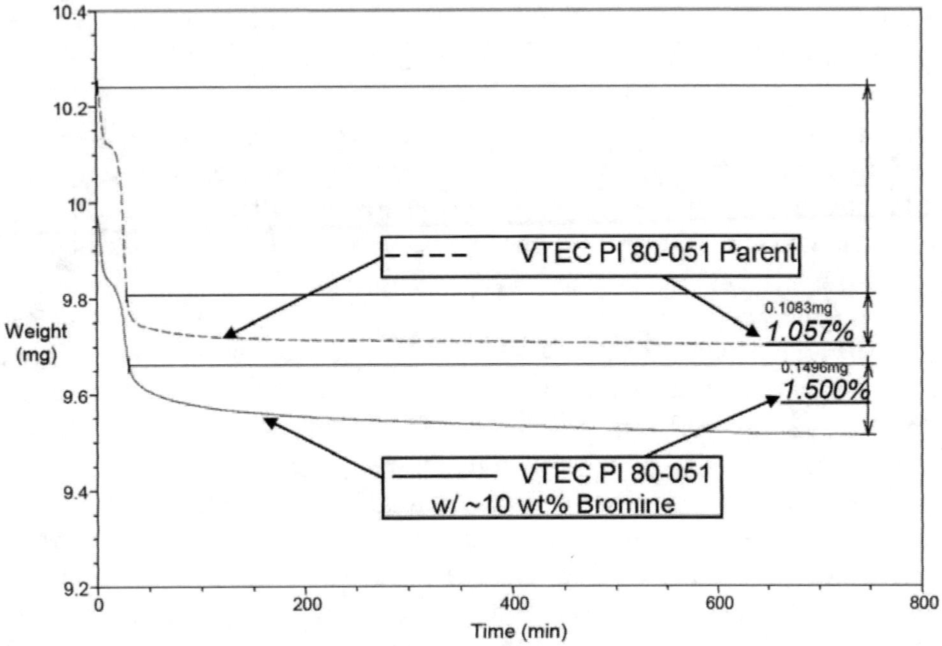

Figure 2. TGA composite of VTEC 80-051 Parent and VTEC 80-051 w/ ~10 wt% bromine at 300°C for 12 hours.

Several polymer blends were created by using VTEC polyimide as the base polymer and blending it with the following polymers: poly[bis((trimethylsilyl)methyl)benzimidazole], PBI-TMS and poly(ethylene glycol), PEG. In some cases, the INL was able to blend a large amount of secondary polymer, like 35 wt% of PEG into the parent VTEC without deleterious effects. This is significant to show the stability of the parent VTEC polymers even blended with known unstable polymers at high temperature (> 250 °C). Table 1 shows some of these VTEC blends along with Kapton®. Remarkably, 50 wt% PBI-TMS can be included in these blends, and they show the best gas throughput when compared to the parent polymers. However, the VTEC-PEG blend showed a reduction in permeability for all of the gases tested.

Table 1. Pure Gas Data Collected at 30°C.

Polymer	Permeability (barrers)[a]						Selectivity α	
	H_2	Ar	N_2	O_2	CH_4	CO_2	CO_2/CH_4	H_2/CO_2
VTEC PI 80-051 w/ ~50wt% PBI-TMS	12.05	1.077	0.907	1.392	0.849	3.987	4.7	3.0
VTEC PI 1388 w/ ~50wt% PBI-TMS	6.655	0.156	0.064	0.442	0.026	1.896	72.9	3.5
VTEC PI 80-051 w/ ~35wt% PEG	1.392	0.017	0.009	0.07	0.006	0.25	41.7	5.6
Kapton®-HN (DuPont)	1.54	*Not tested*	*Not tested*	*Not tested*	0.0045	0.32	71.1	4.8

[a]barrers - (10^{-10})((cm^3 (STP) x cm)/(cm^2 x sec x cmHg))

A critical variable when working with the glassy polymers is their moisture content. Moisture content of the polymer solutions plays a major role with the gas selectivity, especially when the gas permeation rates are low. It has been found that water entrapped within the polymer matrix (either as hydration molecules attached to salts in the polymer, left over solvent, or physisorbed) can also cause the polymer to change dramatically (Table 2). This is evident with the time lag (gas permeability) values determined in the gas permeation experiments. In some cases, the same membrane gave different gas permeation values simply because the membrane was exposed to a humid environment between tests. INL has found that the presence of moisture in the gas stream or in the polymer generally decreases the gas selectivity, while a scrupulously dried membrane will show better performance. Upon discovery that moisture from the environment was being absorbed into the membranes, additional effort was made to dry the films prior to testing. These efforts comprised drying at 150°C for longer periods (24-48 hours). In addition, the membranes were sometimes further dried at 250°C for another 24 hours, such as for VTEC PI 08-051 w/ 10 wt% LiCl. After the drying, the membrane was then quickly transferred to the gas testing apparatus and placed under vacuum for the gas permeation evaluation using dry gas streams.

Table 2. Pure Gas Data Collected at 30 °C

Polymer	Permeability (barrers)[a]						Selectivity α	
	H_2	Ar	N_2	O_2	CH_4	CO_2	H_2/CO_2	H_2/CH_4
VTEC PI 80-051w/~10 wt% LiCl (wet)	4.20	0.35	0.11	0.20	1.01	1.38	3.0	1.4
VTEC PI 80-051w/~10 wt% LiCl (dry)	4.30	0.04	0.03	0.18	0.03	0.58	7.4	19.3

[a]barrers - $(10^{-10})((cm^3\ (STP) \times cm)/(cm^2 \times sec \times cmHg))$

Presence of molecular water in the polymer's void volume has been validated through Positron Annihilation Lifetime (PAL) spectroscopy for VTEC 80-051 (Table 3). From our PALS results, the differences between the lifetimes and intensities of the spectra for unheated and heated samples are relatively small. Since the microporosity of these samples is low compared to other open volume polymers, their long lifetime and intensity values are low, thus their values have relatively high uncertainty. In the case of low statistics (less than 1 million counts in the spectrum), the software program may act in the unstable way. Therefore, the best indicator of any possible change in the structure is the product of both (intensity and lifetime). Since the free volume fraction is calculated based on both—lifetime (pore size) and the intensity (pore frequency, rate)—it is the most reliable value to consider while comparing the unheated and heated spectra of polymers. It is interesting to note that the free volume fraction for VTEC polymers increased after the polymer was heated. This may be caused by the fact that more positrons could be trapped inside the volume of the polymers due to the release of the bound water, since the intensity (I_3) has increased significantly.

Table 3. Estimation of Free Volume from PAL measurements

Polymer	τ_L	τ_L error	I_L %	I_L error	Pore Radius [Å]	Pore Vol. [Å^3]	Free Volume Fraction[c]
VTEC PI 80-051[a]	2.15	0.67	2.6	1.6	3.01 (2.33-3.55)	114 (53-187)	0.53 (0.10-1.42)
VTEC PI 80-051[b]	1.8	0.26	4.42	0.86	2.69 (2.4-2.93)	82 (58-105)	0.65 (0.37-1.00)

[a]Normal atmospheric humidity exposure (20-40% relative humidity)
[b]Heated at 150°C for 1 hour
[c]Free Volume Fraction calculated according to the formula: $f_v(\%)=0.0018*I_3(\%)*\langle 4/3\pi R^3\rangle$

Summary

The N-substituted polybenzimidazole (PBI) post-polymerization polymer modification route provides a straightforward synthetic method that can be carried out at room temperature and gives reasonable yields while making materials that are more amenable to processing. However, our gas testing data for N-substituted PBI polymers show that the gas permeabilities were higher, and the gas separations (like H_2/CO_2) were not as good as the parent PBI material. Therefore, VTEC polyimides along with N-substituted PBI and other polymer systems were developed into membranes that can provide better gas separation. Several polymer blends were created by using VTEC polyimide as the base polymer and blending it with other polymers. These blends have very different flux properties than the starting polymeric films. Finally, a notable problem when using glassy polymers is their moisture content, especially when making blends with other compounds. It has been found that water entrapped within the polymer matrix (left over from the solvent, or physisorbed) can also cause the polymer's fluxes and selectivities to change dramatically. Additionally, the presence of molecular water in the polymer's void volume has been validated through Positron Annihilation Lifetime (PAL) spectroscopy.

Acknowledgements

The work described in this paper was supported by the United States Department of Energy's National Energy Technology Laboratory through Contract DE-AC07-99ID13727 and the Internal Research and Development (IR/D) Program at the National Science Foundation.

References

1. Vogel, H.; Marvel, C. S. *J. Polym. Sci.* **1961**, *50*, 511.
2. (a) Marvel, C. J.; Vogel H. A. Re. 26065, Jul 19, 1966. (b) White, D. M. U.S. Patent 3,408,366, Oct 29, 1968. (c) Chenevey, E. C.; Conciatori, A. B. U.S. Patent 3,549,603, Dec 22, 1970. (d) Sayigh, A. A. R.; Tucker, B. W.; Ulrich, H. U.S. Patent 3,708,439, Jan 2, 1973. (e) Sheratte, M. B. U.S. Patent 4,154,919, May 15, 1979. (f) Choe, E. W. U.S. Patent 4,312,976, Jan 26, 1982. (g) Conciatori, A. B.; Choe, E. W.; Hall, H. K. U.S. Patent 4,483,977, Nov 20, 1984.
3. (a) Barrer, R.M. *Trans. Faraday Soc.* **1939**, *35*, 628. (b) Wijmans, J.G.; Baker, R.W. *J. Mem. Sci.* **1995**, *107*, 1. (c) Rogers, C.; Meyer, J.A.; Stannett, V.; Szwarc, M. Studies In The Gas And Vapor Permeability Of Plastic Films And Coated Papers, Parts I And II. TAPPI **1956**, 39, 737. (d) van Amerongen, G.J. *J. App. Physics* **1946**, *17*, 972.
4. Klaehn, J. R.; Orme, C. J. Luther, T. A.; Jones, M. G.; Wertsching, A. K.; Peterson, E. S., "Soluble N-Substituted Organosilane Polybenzimidazoles," *Macromolecules*, **2007**, *40*, 7487.
5. (a) Gieselman, M. B.; Reynolds, J. R. *Macromolecules* **1992**, *25*, 4832. (b) Gieselman, M. B.; Reynolds, J. R. *Macromolecules* **1993**, *26*, 5633.

CARBON DIOXIDE SELECTIVE SUPPORTED IONIC LIQUID MEMBRANES: THE EFFECT OF CONTAMINANTS

David R. Luebke, Jeffery B. Ilconich(Parsons/RDS), Christina R. Myers, and Henry W. Pennline
United States Department of Energy, National Energy Technology Laboratory, Pittsburgh, PA

Introduction

The integrated gasification combined cycle (IGCC) is widely viewed as a promising technology for the large scale production of energy in a carbon constrained world. The cycle, which can include gasification, contaminant removal, water-gas shift, CO_2 capture and compression, and combustion of the reduced-carbon fuel gas in a turbine, has significant thermal efficiency advantages over conventional combustion technologies. A CO_2 selective membrane capable of maintaining performance at conditions approaching those of low temperature water-gas shift (260°C) could facilitate the production of carbon-neutral energy by simultaneously driving the shift reaction to completion and concentrating CO_2 for sequestration.

Supported ionic liquid membranes (SILMs) have been previously evaluated for this application. Early attempts suffered from support failures due to plasticization (1), but with improved supports, SILMs were determined to be physically and chemically stable to temperatures in excess of 300°C. These membranes were based on ionic liquids which interacted physically with CO_2 and diminished considerably in selectivity at higher temperatures. To alleviate the problem of diminishing selectivity, the original ionic liquids were replaced with ionic liquids able to form chemical complexes with CO_2. These complexing ionic liquid membranes have a local maximum in selectivity which is observed at increasing temperatures for more stable complexes. Similar results have been published by Matsuyama and coworkers (2). Efforts are currently underway to develop ionic liquids with selectivity maxima at temperatures greater than 75°C, the best result to date, but other practical concerns must also be addressed if the membrane is to be function under water-gas shift conditions.

A CO_2 selective membrane must function not only at high temperature, but also in the presence of all the reactants and contaminants likely to be present in coal-derived fuel gas, including water, CO, CH_4, COS, and H_2S. A study has been undertaken which examines the effects of some of these gases on SILMs.

Experimental

Porous films 25mm in diameter and approximately 152µm in thickness of amphoteric nylon-6,6 (Biodyne®A, Pall Corporation) or polysulfone (Tuffryn®, Pall Corporation) (PSF) were used as supports for the ionic liquid membranes. These substrates have 75-85% void volume with nominal pore sizes of 0.2µm. The supported ionic liquid membranes were made by placing supports in a container and depositing ionic liquid on top of the membrane with a pipette. Enough ionic liquid was added to completely cover the surface of the substrate, and the membrane was allowed to absorb the ionic liquid for at least eight hours. The SILMs were then removed from the container and excess ionic liquid was removed by blotting. Finished membranes were placed on the permeate side of a Millipore® filter holder and a polymer substrate identical to the one used to prepare the membrane was placed against the membrane on the feed side to reduce mechanical stress. The active sample area for the holder is 2.2 cm^2.

Steady state performance measurements were made in a flow system in which the permeate and retentate gas compositions were measured using a HP 5890 gas chromatograph with twin TCD detectors and Alltech Hayesep D 100/120 packed columns. Entering gas flows were controlled with rotameters and exiting flows were measured with a digital, bubble flow meter. Temperature was measured by a Type K thermocouple in contact with the surface of the testing cell and controlled using a Lindberg/Blue M tube furnace with integrated PID controller. Feed rate to the sample cell was 2-7 ml/min. Feed gases were purchased premixed and varied according to the experiment: CO_2 (2.00%), CH_4 (2.50%), and argon (balance) with a feed rate of 2 ml/min; CO_2 (19.95%), H_2 (20.01%), H_2S (0.888%) and argon (balance) with a feed rate of 7 ml/min; CO_2 (20.00%), H_2 (19.90%), CO (5.00%), and argon (balance) with a feed rate of 7 ml/min; and CO_2 (19.97%), H_2 (20.02%), H_2O (97ppm) and argon (balance) with a feed rate of 7 ml/min. A sweep of argon (99.999%) was passed over the opposing (permeate) membrane face at a flow of 6-10 ml/min. The total pressure was approximately 108 kPa for the feed and the permeate pressure was less then 105 kPa. After introduction of a feed, the system was allowed to reach steady state and data were then recorded for approximately 4 hours before ramping to the next temperature. Temperature changes were carried out at a rate of 1°C/min for a temperature difference equal to or less than 20°C and 1°C/ 2 min for temperature differences greater than 20°C.

The ionic liquid tested, 1-n-hexyl-3-methylimidazolium bis(trifluoromethanesulfonyl)imide ([HMIM][Tf2N]), was synthesized by the Brennecke research group at the University of Notre Dame using standard procedures (3,4). Impurity levels of halide (Br-) were measured using an Oakton Ion 510 meter with Cole-Palmer Ion Specific Probes (27502-05) and were less than 10 ppm.

Results and Discussion

Membranes consisting of the ionic liquid [HMIM][Tf2N] supported on porous nylon-6,6 had CO_2 permeabilities increasing from 498 to 1165 Barrer over the temperature range 37-300°C with a reasonably strong Arrhenius dependency (R^2=0.9581). The CO_2/H_2 selectivity, as determined by the ratio of the permeabilities of the two gases when both are simultaneously diffusing through the membrane, varied from 9.0 to 1.3 over the same temperature range and showed a somewhat stronger Arrhenius correlation (R^2=0.9687). The data are shown in Figure 1.

In the presence of 5.00% CO, membranes consisting of the ionic liquid [HMIM][Tf2N] supported on porous nylon-6,6 had CO_2/H_2 selectivity decreasing from 7.6 to 2.4 over the temperature range 37-125°C with a strong Arrhenius dependency (R^2=0.9995). The data are shown in Figure 2. The CO concentration is between that of shifted and unshifted fuel gas. Based on these results it may be suggested that the 1% CO typical of fully shifted fuel gas would have little or no effect on the membrane performance, but further tests will be necessary to determine whether or not the higher concentrations found in unshifted fuel gas will present more of a problem.

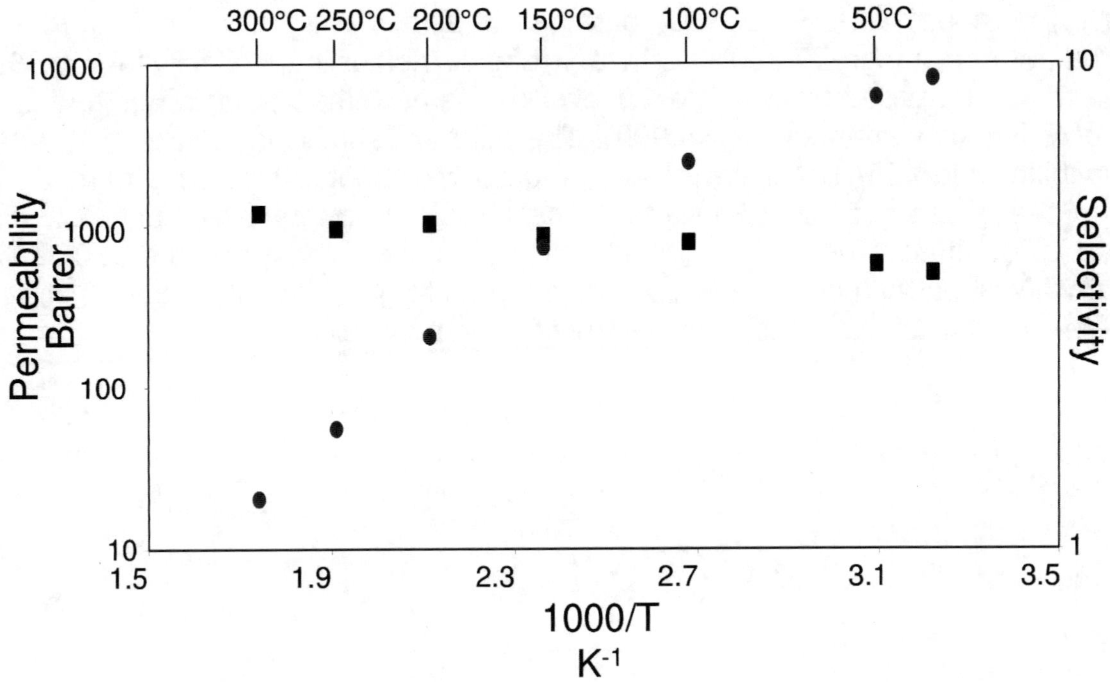

Figure 1. CO_2 permeability (■) and CO_2/H_2 selectivity (•) of a membrane composed of the ionic liquid [HMIM][Tf2N] supported on porous nylon-6,6 as a function of temperature.

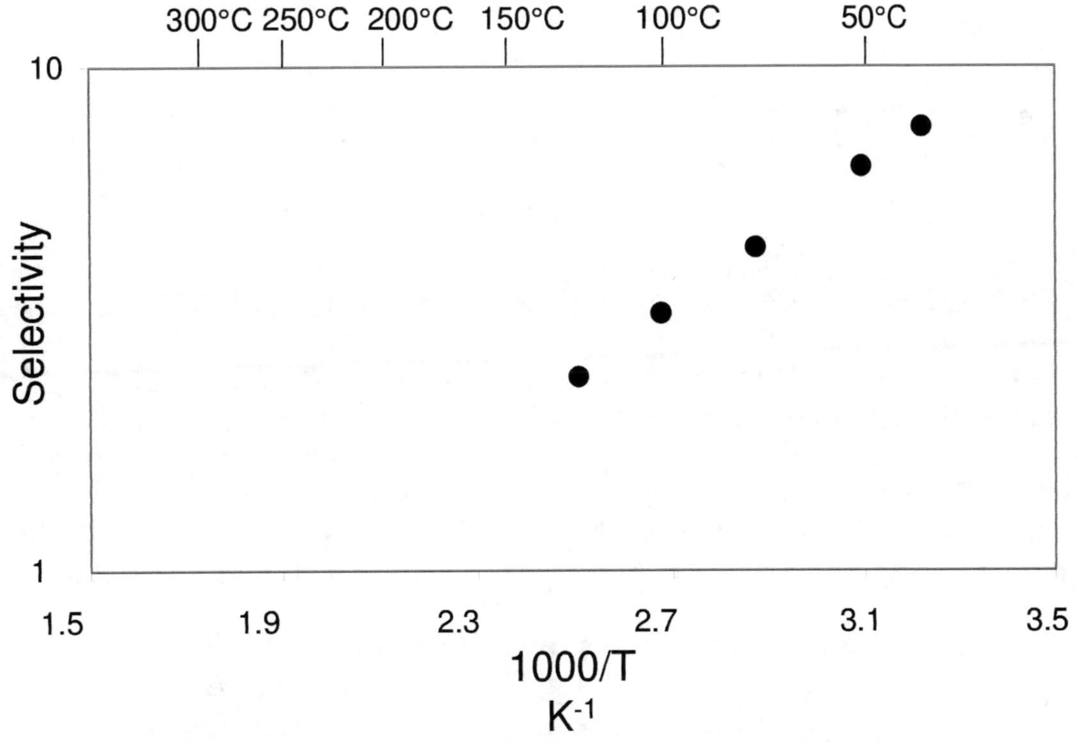

Figure 2. CO_2/H_2 selectivity (•) of a membrane composed of the ionic liquid [HMIM][Tf2N] supported on porous nylon-6,6 as a function of temperature in the presence of 5% CO.

In the presence of 0.89% H_2S, membranes consisting of the ionic liquid [HMIM][Tf2N] supported on porous nylon-6,6 had CO_2 permeabilities increasing from 400 to 1030 Barrer over the temperature range 37-200°C with a strong Arrhenius dependency (R^2=0.969). The CO_2/H_2 selectivity decreased from 7.2 to 1.1 over the same temperature range and showed a very strong Arrhenius correlation (R^2=0.999). The data are shown in Figure 3. At 37°C, both CO_2 permeability and CO_2/H_2 selectivity are reduced by about 20 percent from the values observed in the absence of contaminants. The change is appreciable, but not a serious degradation in the effectiveness of the membrane. It should also be noted that 0.89% H_2S is a higher concentration than would be encountered in most gasification schemes since a sulfur removal step would most likely occur before the CO_2 membrane.

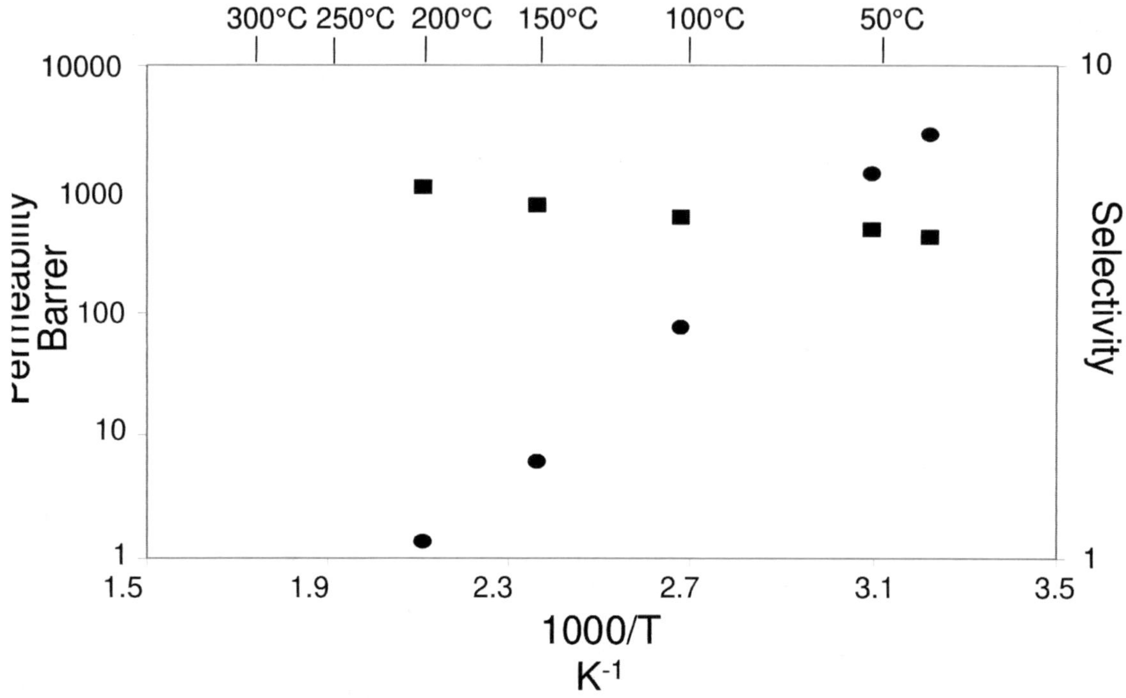

Figure 3. CO_2 permeability (■) and CO_2/H_2 selectivity (•) of a membrane composed of the ionic liquid [HMIM][Tf2N] supported on porous nylon-6,6 as a function of temperature in the presence of 0.89% H_2S.

Information about the possible role of another fuel gas component may be gleaned from the results of experiments evaluating the applicability of the membrane to natural gas sweetening. The membranes had been tested for CO_2/CH_4 separation performance. Comparing the CO_2 permeabilities measured in the presence of 2.5% CH_4 to the permeabilities in the absence of methane, an ideal selectivity may be calculated. As with H_2S, the presence of methane results in appreciable but not critical reduction in both permeability and selectivity.

Table 1. CO_2 permeability of a membrane composed of the ionic liquid [HMIM][Tf2N] supported on porous nylon-6,6 in the presence of 2.5% CH_4.

	37°C	60°C
CO_2 Permeability (Barrer)	378	416
CO2/H2 Selectivity (Ideal)	6.8	4.3

Membranes consisting of the ionic liquid [HMIM][Tf2N] supported on porous PSF had been previously studied. The support was abandoned because it lacked the thermal stability required to perform at water-gas shift reaction conditions. Membranes fabricated using the PSF support showed higher permeability and slightly lower selectivity than those which employ the nylon-6,6 support owing to the lesser contribution of gas diffusion through the polymer rather than the ionic liquid in the nylon-6,6 membrane. The support should have little bearing on contaminant effects since those will be confined to alternation of the ionic liquid rather than the support. A comparison of the performance of this membrane in the presence and absence of 97 ppm water is shown in Table 2. The changes in performance are small, leading to the conclusion that at low concentrations water does not play a significant role in membrane performance.

Table 2. CO_2 permeability and CO_2/H_2 selectivity of a membrane composed of the ionic liquid [HMIM][Tf2N] supported on porous PSF under dry conditions and in the presence of 97ppm H_2O.

	Dry		Wet	
Temperature (°C)	37	100	37	100
CO_2 Permeability (Barrer)	840	1040	742	1060
CO_2/H_2 Selectivity	7.0	3.0	6.8	2.9

Examining the results above, it is possible to suggest that in the short term common fuel gas contaminants will not seriously deteriorate the performance of non-complexing ionic liquid membranes. Additional tests are required to determine what the long term effect of these contaminants may be, to examine how performance changes in the presence of multiple contaminants, to conduct investigations under realistic pressures, and to explore the effect of contaminants on chemically complexing ionic liquids.

Acknowledgements

The authors gratefully acknowledge the contributions of the Joan Brennecke and Ed Maginn research groups at the University of Notre Dame. Professors Brennecke and Maginn have been invaluable both in educating us with respect to ionic liquids and in providing the ionic liquid samples and characterization data critical to this work. We thank them for their efforts and look forward to further fruitful collaboration in the future.

Disclaimer

References

1. J. Ilconich, C. Myers, H. Pennline, D. Luebke, (2007) "Experimental investigation of supported ionic liquid membranes for CO_2/He separation at temperatures up to 125°C," *Journal of Membrane Science,* 298, pp. 41-47.

2. S. Hanioka, T. Maruyama, T. Sotani, M. Teramoto, H. Matsuyama, K. Nakashima, M. Hanaki, F. Kubota, M. Goto, (2007), "CO2 separation facilitated by task-specific ionic liquids using a supported liquid membrane," Journal of Membrane Science, doi:10.1016/j.memsci.2008.01.029.
3. J.M. Crosthwaite, S.N.V.K. Aki, E.J. Maginn, J.F. Brennecke, (2004) "Liquid phase behavior of imidazolium-based ionic liquids with alcohol," *Journal of Physical Chemistry B,* 108, p. 5113.
4. J.M. Crosthwaite, S.N.V.K. Aki, E.J. Maginn, J.F. Brennecke, (2005) "Liquid phase behavior of imidazolium-based ionic liquids with alcohols: effect of hydrogen bonding and non-polar interactions," *Fluid Phase Equilib.*, 228, p 303.

22d Theoretical Predictions of Hydrogen Permeability in Pd-Based Membranes

Jennifer Wilcox, Energy Resources Engineering, Stanford University, 367 Panama Street, Green Earth Sciences 065, Stanford, CA 94305

The present work deals with the study of palladium-silver (PdAg) and palladium-gold (PdAu) binary alloys over a broad range of temperatures and alloy compositions using density functional theory (DFT) to find out possible conditions where the solubility, diffusivity, and permeability of hydrogen (H) is significantly higher than that of pure palladium (Pd). Several alloy structures, such as Pd(100-x)Ag(x) with x=14.81, 25.93, 37.04 and 48.51, Pd(100-x)Au(x) with x=14.81, 25.93, 37.04, and Pd(100-x)Cu(x) with x=25.93 and 48.51 have been considered. The lattice constants of these structures were optimized using DFT, and relaxed structures were used for the estimation of binding energy. It was found that the solubility of H in PdAg is higher than pure Pd with a maximum at approximately 30%Ag at 456K. Also, the solubility of PdAu alloys was higher than the pure Pd with a maximum at about 20%Au with solubility 12 times higher than that of pure Pd. It was found that for a 3.7% H concentration in a PdAg alloy, a cell expansion of 0.15-0.2% occurs, which if ignored may affect the individual binding energy of the O-site by approximately 3.56% and affect the predicted solubility by approximately 11.8%.

Kinetic Monte Carlo simulations (KMC) were carried out using predicted rate constants for hydrogen hopping throughout simulated Pd-based alloy lattices. DFT has been employed to obtain octahedral (O-), tetrahedral (T-) and transition state (TS-) site energetics as a function of local alloy composition for several PdAg and PdAu alloys with compositions in supercells of X=14.18, 25.93, 37.07 and 48.15% with the nearest (NN) and next-nearest neighbors (NNN) varied over the entire range of compositions. The estimates were then used to obtain a model relating O-, T-, and TS-energy of a given site with NN (X), NNN(X), and the lattice constant. The first passage approach combined with KMC simulations was used for the H diffusion coefficient predictions. It was found that the diffusion coefficient of H in PdAg alloy decreases with increasing Ag and increases with increasing temperature, matching closely with the experimental results reported in the literature. The calculated permeabilities of H in these novel binary alloys obtained from both diffusivity and solubility predictions were found to have a maximum at ~20% Ag and ~12% Au, which agree well to experimental predictions. Specifically, the permeability of H in PdAg alloy with ~20% Ag at 456K is 3-4 times that of pure Pd, while the PdAu alloy at 12% Au is 4-5 times that of pure Pd at 456 K.

Post-Combustion Dry Phase CO_2 Capture with a Charged Polymeric Membrane

Gordon Nangmenyi[1], Michael Heiber[2] and James Economy[1]

(1)Dept. of Materials Sci. & Eng., University of Illinois, 1304 W. Green St. Urbana, IL 61801. (2) University of AkronDepartment of Polymer Science 170 University Circle, Goodyear Polymer Center Rm. 429, Akron, OH 44325.

Abstract

Power plants, particularly coal-fired plants, are the largest single point source of CO_2 emissions, owning 39.4% of the CO_2 released into the atmosphere each year. Research to efficiently capture CO_2 in the dry gas state from power plants has received great interest in recent years. A novel microporous polymeric based material composed of poly(ethyleneimine) and epoxy on a non-woven fiberglass support [PEIM/EP] has been specifically developed to address post-combustion CO_2 capture from power plants. The material system exhibits high thermal stability in air (up to 200°C) and high CO_2 adsorption capacity (440 mg CO_2 [g PEIM/EP]$^{-1}$) at 28°C and 85% high relative humidity, near the humidity level a typical coal-fired power plant process. The adsorption profile of increasing partial pressures of CO_2 up to 98.6 ± 0.2 kPa for the material system strongly fit the Langmuir isotherm. It is also capable of multiple regeneration and reuse with temperature swing adsorption (TSA) at 140°C. Synthesis and characterization of the material including surface area, pore volume and selectivity will be discussed.

Introduction

In efforts to mitigate the release of CO_2 from the flue gas exhaust of power plants, which are the largest single source emitters of CO_2 (39.4% of the CO_2 released into the atmosphere per year), the most widely accepted post-combustion capture technologies have used chemically driven absorption processes which rely on alkaline solvents. [1] Though these technologies have displayed excellent results, with some achieving greater than 90% removal of CO_2 from the effluent flue gas, they have inherent disadvantages including limited to high heat of reaction, volatility, rate of degradation and waste generation as well as high energy consumption for regeneration. [2; 3] Polymeric membrane separation and capture processes offer key advantages over solvent based systems which include 1) lower energy requirement of separation after capture CO_2 capture and 2) lower equipment and maintenance costs.

Researchers have demonstrated the utility of poly(ethyleneimine) in acid gas adsorption. [4,5] Our approach for capturing CO_2 from a power plant's effluent flue gas utilizes a fibrous adsorbent composed of a mixture of polyethylenimine cross-linked with epoxy on a non-woven fiberglass substrate [PEIM/EP]. The PEIM/EP fibrous adsorbent displays the properties for an ideal adsorbent for CO_2: [6]

(1) High selectivity for CO_2.

(2) High adsorption capacity for CO_2.

(3) Good mechanical strength.

(4) Adequate adsorption / desorption kinetics.

(5) Stable adsorption capacity after cycles.

The micro and mesopores formed from coating the non-woven fiberglass with poly(ethyleneimine) and epoxy enable the CO_2 adsorption to be driven by chemical and physical adsorption processes. The amine end groups of the poly(ethyleneimine) have an affinity for acidic gases and are selective for CO_2 as well as the other acidic gaseous components such as NO_x and SO_x, although the SO_x is removed in the prior desulfurization step in the standard coal-fired power plant process. The PEIM/EP fibrous adsorbent displays strong CO_2 adsorption and desorption at or above 140°C.

Experimental

<u>Adsorbent Preparation</u>

The fibrous adsorbent, PEIM/EP is prepared through a wet impregnation method. The typical process uses an epoxy (Huntsman Advanced Materials) and polyethylenimine (PEI, Mw=25,000, water free, Aldrich Co.). The epoxy is prepared with Epocast 35-A and Hardener 927 in a 4:1 weight ratio. A 9:1 ratio of PEI and epoxy was dissolved in 6.3 g acetone per gram of polymer mixture under stirring for 10 minutes at room temperature. After the solution was completely dissolved, 0.3 mg of non-woven fiberglass mat with 7% PVA (poly(vinyl) alcohol) binder (Craneglas ®230, from Crane Nonwovens, Inc.) and a fiberglass diameter of 6.5 μm was dipped into the solution until completely wetted and dried in air for 12 h at room temperature. The samples were cured by heating at 100°C for 5 h in flowing N_2.

<u>Characterization</u>

A thermal gravimetric analyzer (TGA 951, TA Instruments) was employed to measure the mass change of the PEI/EP fibrous adsorbent over time at varying CO_2 partial pressures. Tylan FC280 mass flow controllers were used to regulate the CO_2 and N_2 flow while an Omegaette HH314 relative humidity meter was used to track the moisture level of the experimental environment. Scanning electron microscopy was used to observe the morphology of the PEIM/epoxy fibrous adsorbent.

<u>Adsorption</u>

Practical use of the PEI/epoxy fibrous adsorbent material requires stability in a saturated water vapor environment. A water adsorption test was performed to assess the crosslinking density of the PEI/EP fibrous adsorbent at varying polymer to epoxy ratios. Prior to the start of the experiment, adsorbed moisture was removed from the PEI/EP fibrous adsorbent by heating in flowing N_2 at 120°C for 6 h. The material was weighed and then placed in a closed water vessel. After three days, the material was heated in flowing N_2 at 120°C and reweighed. The adsorption capacity for water was determined by calculating the difference before and after water adsorption.

CO_2 adsorption experiments were carried out in the experimental set up shown in figure 1. The set up consists of a horizontal TGA (TGA 951, TA Instruments), mass flow controllers, humidifiers and relative humidity meters. First, the PEIM/EP fibrous adsorbent was heated to 95°C for 1 hour under N_2 to release any adsorbed moisture. Second, the dry N_2 is piped through a deionized water (DIW) column that introduces water vapor to the membrane. After equilibrium is reached, the gas flow is switched from N_2 to CO_2 which is also bubbled through a separate column of DIW to maintain the same level of water vapor. The mass gained at this step is assumed to be the amount of CO_2 adsorbed. Lastly, desorption of the adsorbent is performed at 105°C for 1.5 h. In each test, the temperature was in the range of 25-28°C and the relative humidity ranged between 64-82%.

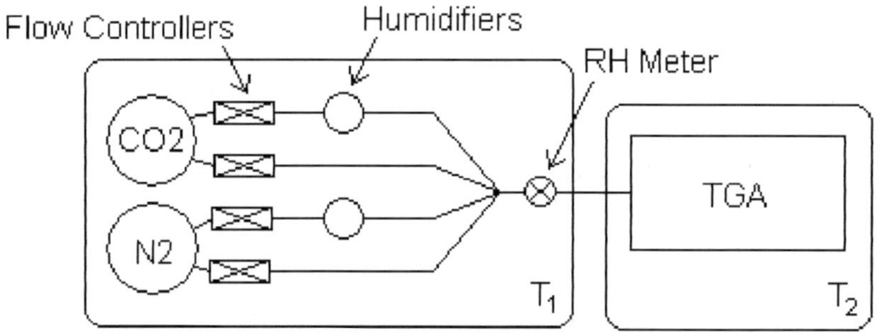

Figure 1: Experimental apparatus consists of 4 dual valve mass controllers, 2 humidifiers, relative humidity meter and horizontal TGA to measure the mass basis of adsorption.

Results and Discussion

It was discovered that CO_2 adsorption capacity improved with humidity. Figure 2 displays the typical adsorption curve generated for the PEIM/EP fibrous adsorbent.

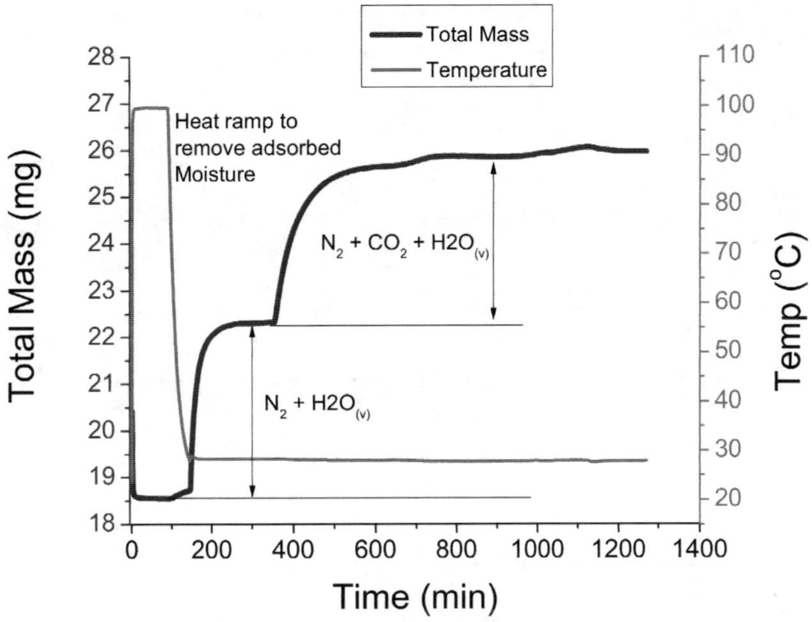

Figure 2: Typical CO_2 adsorption curve.

CO_2 adsorption capacity was calculated at varying CO_2 partial pressures. The PEIM/EP displayed a maximum CO_2 adsorption capacity of 440 mg CO_2 [g-PEIM/EP]$^{-1}$ with a relative humidity of 85% at 28°C. Figure 3 displays the CO_2 adsorption capacity at increasing CO_2 partial pressures. The CO_2 adsorption capacity also increased with increasing weight ratio of PEI to epoxy. Maximum CO_2 capacity was observed at a weight ratio of 95/5 though there was a loss in crosslinking density. CO_2 adsorption was further increased with the presence of micro-silica beads within the polymer matrix by greater than 50% for a weight ratio 9:1 PEIM/EP.

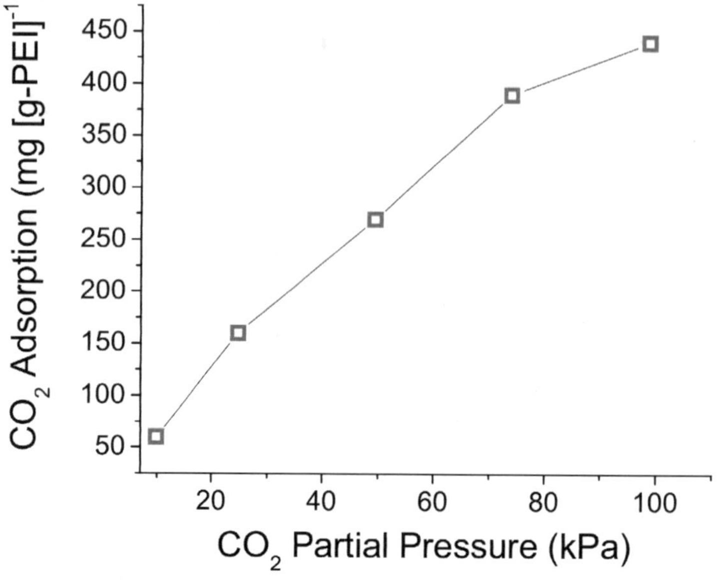

Figure 3: CO_2 adsorption capacity versus CO_2 partial pressure.

References

1. **Sharp, E., Parsons, S., and Jefferson, B.** *Seasonal Variations in Natural Organic Matter and its Impact on Coagulation in Water Treatment.* 2006, Science of the Total Environment, Vol. 363, pp. 183-194.

2. **Bolland, O., Lindeberg, E., et al.** *Power Generation with CO2 Sequestration - Research and Development Needs.* Trondheim : Sinteff Energy Research, 2000. 17X236.01.

3. **Verseeg, G. and Van Swaaij, W.** *On the kinetics between CO2 and alkanolamines both in aqueous and non-aqueous solutions – II. Tertiary amines.* 1988, Chemical Engineering Science, Vol. 43, pp. 587-591.

4. **Xu, X. et al.** *Preparation and Characterization of Novel CO_2 "Molecular Basket" Adsorbents Based on Polymer-modified Mesoporous Molecular Sieve MSCM-41.* 2003, Microporous and Mesoporous Materials. Vol. 62, pp. 29-45.

5. **Andreopoulos, A.G. and Economy.** *Design of High-Capacity, Reversible Polymeric Adsorbents for Acidic Gases.* 1991, Polymers for Advanced Technologies, Vol. 2, pp. 87-91.

6. **Mata, V. and Rodrigues, A.E.** *Adsorption of Carbon Dioxide at High Temperature – a Review.* **Yong, Z.,** 2002, Separation and Purification Technology, Vol. 26, pp. 195-205.

CAPTURE OF CO_2 FROM COMBUSTION FLUE GAS USING THE CARBOZYME LIQUID MEMBRANE PERMEATOR – RESULTS OF PILOT-SCALE TESTING

Melanie D. Jensen and Jason D. Laumb
Energy & Environmental Research Center
Grand Forks, ND

Robert M. Cowan, David A. Smith, and Michael C. Trachtenberg
Carbozyme, Inc.
Monmouth Junction, NJ

INTRODUCTION

Carbozyme, Inc., has developed a biomimetic technology that appears to offer significant cost and performance advantages over monoethanolamine (MEA) scrubbing systems for capture of CO_2 from combustion flue gases. The Carbozyme technology is based on separation of CO_2 from a mixed-gas stream as a bicarbonate, whose reaction is catalyzed by carbonic anhydrase (CA), an enzyme found in plants and animals. The process uses a hollow fiber-based permeator design that features a layer of CA in a thin liquid film, called a contained liquid membrane (CLM), contained between two polymer membranes.

A $4.8 million cooperative agreement with the U.S. Department of Energy National Energy Technology Laboratory's Carbon Sequestration Program has been secured to further develop and scale up the Carbozyme process for the capture of CO_2 from power plant flue gas. A multistep approach has been adopted to further develop the Carbozyme process, including the following activities:

- The composition of the flue gases most likely to be encountered at coal-fired facilities in the nation's power generation fleet is being estimated based on coal type, burner configuration, and downstream air pollution control devices.

- A prototype CLM module is being fabricated and installed downstream of the pollution control equipment on a 42.2-MJ/h (40,000-Btu/h) combustion and environmental process simulator (CEPS) system at the Energy & Environmental Research Center (EERC).

- Four, 250-h tests on the CEPS will evaluate the ability of the Carbozyme CLM to capture CO_2 from flue gases generated during the combustion of natural gas, lignite, subbituminous coal, and bituminous coal at both steady-state and upset conditions. Longer-term (~2000-h) testing is planned. The resulting data will be used to validate and further scale up the technology 100-fold.

- The scaled-up Carbozyme CLM system will be installed as a multimodule, skid-mounted system on a 527.5–MJ/h (500,000-Btu/h) combustion system at the EERC where it will be evaluated during a 1-wk, 50h test. Data collected will be used to determine CO_2 capture efficiency and estimate capture costs using this technique.

As is the case with virtually all other CO_2 capture technologies, a pretreatment system is needed to polish the flue gases from diverse feedstocks to remove acid gases, acid aerosols particulate, and ionized mercury that could interfere with operation of the module. Removing these species will allow the Carbozyme process to be applied as broadly as possible to the fleet of U.S. coal-fired power plants and other large point sources of CO_2. The first step in determining the most appropriate pretreatment system is to identify the range of combustion flue gas constituents that will be encountered most often. This was the objective of the work that is described in this paper.

APPROACH

It was considered important that the Carbozyme technology apply not only to natural gas-fired facilities (the flue gases of which contain relatively few contaminants) but to as many coal-fired power plants as possible. Ideally, a pretreatment scheme would be identified that would clean all flue gases sufficiently, irrespective of coal type or existing plant pollution control devices, that they would be acceptable for delivery to the Carbozyme permeator. The potential flue gas compositions that would be encountered had to be defined so that an appropriate pretreatment approach could be determined. Direct measurements were not possible because very few power plants will provide detailed compositional data for their flue gases. It was determined that the flue gas compositions would have to be estimated for as broad a range of coals and pollution control devices as possible. The expected flue gas compositions were defined through the following steps:

- The most commonly used coals and their proximate and ultimate analyses were identified.
- The most common plant configurations (i.e., boiler type and attendant pollution control devices) were determined for the entire U.S. power production fleet.
- The compositions of the flue gases produced by the most mined coals when fired in the most common plant configurations were estimated.
- The estimated flue gases were compared to identify a range and median value for each component that could be expected.
- The expected component ranges were compared with the Carbozyme permeator acceptance criteria to determine the cleanup requirements of the pretreatment technology.

METHODS

The largest production mines for the three ranks of coal that are typically burned in power plants in the United States (i.e., bituminous and subbituminous coals and lignites) were identified and are summarized in Table 1. While these coals may not be used exclusively in power plants, the majority of coal that is mined is used for this purpose.

Table 1. Most Mined Bituminous Coal, Subbituminous Coal, and Lignite in the United States

Seam or Mine(s) Name	State
Bituminous Coal	
Pittsburgh No. 8	Pennsylvania and West Virginia
Foidel Creek	Colorado
Sufco	Utah
Century	Ohio
Powder River Basin Subbituminous Coal	
Black Thunder	Wyoming
North Antelope–Eagle Butte	Wyoming
Cordero	Wyoming
Rosebud	Wyoming
Caballo	Wyoming
Buckskin–Belle Ayr–Antelope	Wyoming
Lignite	
Freedom	North Dakota
Falkirk	North Dakota
Big Brown	Texas
Center	North Dakota

The most common power plant configurations were determined using the EERC's U.S. Environmental Protection Agency (EPA) mercury information collection request (ICR) database.[1] The ICR database comprises mercury emission data for select coal-fired electric utilities in the United States. In response to the Clean Air Act Amendments, EPA instituted the ICR to determine if a need existed to regulate mercury emissions from those facilities. The database was compiled and is maintained by the EERC. In addition to coal analyses and mercury emissions, data contained in the database for each power plant include rank of coal fired, plant size, and installed pollution control devices.

The proximate, ultimate, and trace element analyses for the most commonly fired coals within each coal rank were obtained from the U.S. Geological Survey (USGS) database.[2] Many of the coals contained in the database are stored by seam name. To access these coals, the *Keystone Coal Industry Manual*[3] was used to identify the appropriate seam for each coal. The analytical results obtained from the USGS Web site were averaged and the average values input into the Carnegie Mellon Integrated Environmental Control Model (IECM) to obtain an estimate of the flue gas composition that would be produced. The IECM is a computer program that can be used to carry out cost and performance analyses of emission control equipment at coal-fired power plants.[4] A variety of pollution control technologies can be applied to the configuration of the facility that is being modeled.

The most common power plant configurations (representing about 46% of the U.S. power plant fleet) were input into the model for each of the coals in order to obtain flue gas compositions. Each configuration consisted of boiler type and SO_x, NO_x, and particulate control. The IECM does not have calculation loops for hot-side electrostatic precipitators (ESPs) or ESP–fabric filter combinations; thus configurations containing these devices could not be included. The model was run 92 times to obtain flue gas compositions for each of the different coal–power plant configuration combinations. A 500-MW, subcritical, pulverized coal-fired power plant operating at 75% capacity was chosen as the base plant, with IECM default values used as the inputs for the various plant unit operations.

RESULTS AND DISCUSSION

Carbozyme Liquid Membrane Permeator Requirements

Carbozyme has identified flue gas contaminants that can have deleterious impacts on the performance of the CLM module, including particulate, strong oxidants, and water-soluble contaminants such as acid gases.[5] Particulate has the potential to occlude pores in the membrane and, in an extreme case, could clog the bore of the hollow fiber. Strong oxidants or other highly reactive species have the potential to react with the polymer membrane, limiting its useful life. Acid gases and water-soluble oxidized metals will be absorbed into the CLM, which is a lightly alkaline buffer. As the acids and oxidized metals accumulate, the performance of the permeator will decrease because of either a change in pH or because specific ions can act as inhibitors to the enzyme present in the liquid membrane. The liquid membrane permeator acceptance limits are given in Table 2. It should be noted that the acceptance limit for SO_x is quite similar to the SO_x acceptance limit for other CO_2 capture technologies that are commercially available or nearing commercialization. [6]

Estimated Flue Gas Compositions

Table 3 shows the component maxima, minima, and median values for the 92 flue gas composition estimates. It should be emphasized that these are the compositions of the flue gases following the pollution control systems. The goal of these activities was to identify a pretreatment approach that can produce a gas stream that is acceptable to the Carbozyme CLM when the components vary within the ranges shown in Table 3, irrespective of the combustion feedstock. Comparison of Tables 2 and 3 shows that the pretreatment system will have to remove HCl, SO_3, NO_2, and oxidized mercury to meet the permeator acceptance requirements.

Table 2. Carboyzme CLM Acceptance Limits[5]

Contaminant	Acceptance Concentration	
HCl	7 ppmv	8.5 ppmw
HF	7 ppmv	4.6 ppmw
SO_x (SO_2 + SO_3)	7 ppmv	18.6 ppmw
NO[a]	190 ppmv	195 ppmw
NO_2[b]	20 ppmv	21 ppmw
Hg (as Hg^{2+})	4×10^{-6} ppmv	0.03 μg/m^3
Hg (as Hg^{2+}) with 1mM EDTA	$>2\times10^{-3}$ ppmv	>20 μg/m^3

[a] Assumes 3.7% capture (experimentally observed, unpublished data) as HNO_3 and K_i = 100 mM.
[b] Assumes 34.7% capture (experimentally observed, unpublished data) as HNO_3 and K_i = 100 mM.

Table 3. Flue Gas Component Concentration Ranges as Estimated Using the IECM

Flue Gas Component	Bituminous Coals			PRB[1] Subbituminous Coals			Lignites		
	Minimum	Maximum	Median	Minimum	Maximum	Median	Minimum	Maximum	Median
N_2, wt%	65.34	69.83	69.13	62.91	67.69	67.03	62.31	68.75	62.96
O_2, wt%	5.58	5.99	5.92	5.39	5.82	5.74	5.34	5.89	5.41
H_2O, wt%	4.79	10.56	5.45	6.72	12.53	7.53	7.73	14.81	12.28
CO_2, wt%	17.11	18.29	17.87	17.25	19.12	18.31	16.94	18.48	17.66
CO, wt%	0	0	0	0	0	0	0	0	0
HCl, ppmw	0.58	56.1	20.0	0.748	16.9	7.9	0.71	12.4	4.2
SO_2, ppmw	296.2	5137	1324	328.1	2252	850	384.3	2137	397
Sulfuric Acid (equivalent SO_3), ppmw	0.32	19.4	4.4	0.04	0.31	0.3	0.01	1.01	0.25
NO, ppmw	165.9	374.6	200.5	160.7	484	214.0	151.2	459.3	175.6
NO_2, ppmw	13.4	30.2	16.2	13.0	39.0	17.3	12.2	37.1	14.2
NH_3, ppmw	0	2.4	2.4	0	2.5	2.4	0	2.5	2.4
Ar, wt%	1.11	1.19	1.18	1.07	1.15	1.14	1.06	1.14	1.07
Mercury									
Elemental, μg/Nm^3	1.4	4.74	2.73	8.07	13.7	10.32	9.55	22.64	12.66
Oxidized, μg/Nm^3	0.18	11.07	6.36	0.13	4.57	3.09	0.16	7.55	1.78
Total, μg/Nm^3	1.76	15.81	9.08	8.2	18.26	12.89	9.71	30.19	14.54

[1] Powder River Basin.

Because the purpose of the project is to evaluate the Carbozyme liquid membrane permeator technology rather than to develop a new method for removing pollutants from flue gases, technologies that are commercially available were evaluated for their ability to remove the contaminants of concern. Three technologies were considered: condensing heat exchangers, wet ESPs, and wet scrubbers. Wet scrubbing was selected because it was deemed the most likely to scrub out several of the various contaminants and could be relatively easily installed on the combustion test systems. Pozzolanic fly ashes would pose fewer problems in a wet scrubber than in either of the other two technologies. In addition, most utilities understand conventional wet-scrubbing technology and would probably accept this approach if it were to be selected as the polishing technology for full-scale CLM systems. Additional SO_3 removal may be necessary for the most extreme flue gas cases (i.e., from high-sulfur bituminous coals such as from the Century Mine).

Scrubber Design

A packed scrubber tower has been designed for use with the EERC's 42.2-MJ/h CEPS system. For the scale at which this scrubber was designed, random packing best approximates full-scale tray towers. It was decided to construct the scrubber from Plexiglas, randomly pack it with 13-mm ceramic Intalox saddles, and absorb the contaminants using an 11 wt% calcium carbonate slurry as this would provide ample slurry solids loading for adjustment depending upon the actual flue gas composition and the test period requirements. A minimum slurry flow rate was determined stoichiometrically, based on the results of the flue gas composition estimates. The tower design was performed using standard chemical engineering gas–liquid absorption tower column design principles.

The design calculations indicated that the scrubber would have an inside diameter of 10.16 cm (4.0 in.) and an irrigated packed height of 1.3 m (4.28 ft). The fan power requirement is 0.02-hp. The scrubber is currently being fabricated and installed on the CEPS system. Integrated system shakedown will take place as soon as possible, with CLM module testing expected to commence in June 2008.

REFERENCES

1. Laumb, Jason; Jensen, Robert; and Benson, Steven (2000), "Information Collection Request for Mercury: Correlation Analysis of Coal and Power Plant Data," Presented at the Air Quality II Conference, McLean, VA, Sept 2000.

2. U.S. Geological Survey (2007), U.S. Coal Quality Database, National Coal Resources Data System. http://energy.er.usgs.gov/coalqual.htm (accessed Aug 2007).

3. *Keystone Coal Industry Manual* (1992), Chicago, IL, Maclean Hunter Publishing Co.

4. Carnegie Mellon University Department of Engineering and Public Policy (2007), Integrated Environmental Control Model (Feb 2007 version) developed by Carnegie Mellon University Department of Engineering and Public Policy with support from the U.S. Department of Energy National Energy Technology Laboratory. www.iecm-online.com (accessed Aug 2007).

5. Cowan, Robert (2007), Carbozyme, Personal communication with Melanie Jensen, EERC, Aug 16, 2007.

6. Electric Power Research Institute (2007), "Update on Clean Coal Technologies and CO_2 Capture and Storage," presented to the Oregon Public Utility Commission on June 27, 2007, at Salem, OR. www.puc.state.or.us/PUC/meetings/pmemos/2007/062707/OregonPUCCCTCCS62707.ppt.

22g Development of Poly(amidoamine) (PAMAM) Dendrimer Composite Membrane Module for CO_2 Separation

Teruhiko Kai, Shuhong Duan, Firoz Alam Chowdhury, Shingo Kazama, and Yuichi Fujioka. Research Institute of Innovative Technology for the Earth (RITE), 9-2 Kizugawadai, Kizugawa-shi, Kyoto, 619-0292, Japan

Carbon dioxide Capture and Storage (CCS) has been widely accepted as a promising way of mitigating global warming. The first step of CCS is capturing CO_2 from a large stationary CO_2 source such as a power station, where capturing CO_2 accounts for more than 70 % of the total CCS cost. The cost reduction of CO_2 capture is critically important for spreading CCS technology. CO_2 membrane separation is a powerful candidate technology for reducing CO_2 capture costs. Poly(amidoamine) (PAMAM) dendrimer was reported to have excellent CO_2/N_2 selectivity as an immobilized liquid membrane for the isobaric test condition of atmospheric pressure [Kovvali and Sirkar, Ind. Eng. Chem. Res. 40 (2001) 2502]. However, this immobilized liquid membrane had insufficient pressure tolerance for practical use. In previous papers [Kouketsu, Duan, Kai, Kazama and Yamada, J. Memr. Sc. 287 (2007) 51; Kazama, Kai, Kouketsu, Matsui, Yamada, Hoffman and Pennline, Session 30, Proceedings of Pittsburgh Coal Conference, Pittsburgh, USA (2006)], we have prepared stable PAMAM dendrimer composite membranes using chitosan (CTS) as a gutter layer by the In-situ Modification (IM) method. The PAMAM dendrimer composite membrane had high CO_2/N_2 selectivity and can afford a practical tolerance of the pressure difference of 100 kPa. CO_2 permeance of the PAMAM dendrimer composite membrane was reasonably high. However, further improvement for CO_2 permeance was still required. In this presentation, we investigated the effect of the addition of hyaluronic acid (HA) to the CTS gutter layer to improve the CO_2 performance of the composite membrane. The influences of HA addition on the swelling degree in water, the content of PAMAM dendrimer in the gutter layer, and CO_2 separation performance of the composite membrane are discussed. In addition, a novel hydroxyl PAMAM dendrimer was applied to improve the composite membrane performance. The HA addition increased the PAMAM dendrimer content in the gutter layer. A composite membrane prepared with a novel hydroxyl PAMAM dendrimer in the CTS-HA gutter layer exhibited an excellent CO_2/N_2 selectivity of 1,100 and CO_2 permeance of 8.8×10^{-10} m^3 (STP) m^{-2} s^{-1} Pa^{-1}.

32a Session Introduction

Bernard S. Lee, Institute of Gas Technology, 9226 E. Champagne Dr, Sun Lakes, AZ 85248

This session is a panel discussion on massive electricity storage (MES). The session introduction will introduce the topic and invited panelist.

32b Power System Instability

Howard Illian, Energy Mark Inc., 334 Satinwood Ct., North, Buffalo Grove, IL 60089

This presentation will give an overview of why and how a power system becomes unstable or fails when attached to large sources of intermittent power.

32c R&D for Massive Storage Technologies for Solar Power

Brad Roberts, S&C Electricity Storage Association, 5251 Franklin Dr, Franklin, WI 53132

This presentation will describe the need to stimulate development & demonstration in massive storage technologies for solar power.

32d Vanadium Flow Batteries for Wind Generators

Tim Hennessy, VRB Power System Inc., 13955 Bridgeport Rd, Suite 200, Richmond, BC V6V1J6, Canada

This presentation will describe Vanadium Flow Batteries as large storage devices for wind generators.

32e Sodium-Sulfur Batteries for Peak Shaving or Wind Generation

Ali Nourai, American Electric Power, 1 Riverside Plaza, Columbus, OH 43215

This presentation will describe sodium-sulfur batteries for use in peak shaving or wind generation.

32f Massive Electricity Storage Panel Discussion

Bernard S. Lee, Institute of Gas Technology, 9226 E. Champagne Dr, Sun Lakes, AZ 85248 and Dave Gushee, Consultant, Front Royal, VA 22630.

This will be a discussion among the invited panelists on Massive Electricity Storage moderated by the session co-char.

32g Question and Answer Session with Audience

Bernard S. Lee, Institute of Gas Technology, 9226 E. Champagne Dr, Sun Lakes, AZ 85248 and Dave Gushee, Consultant, Front Royal, VA 22630.

This will provide time for the audience to ask the invited panelists questions on Massive Electricity Storage.

59a Introduction and Overview of the Basic Energy Science (BES) Program

Patricia Dehmer, Office of Sciences, U. S. Department of Energy, SC-10/Germantown Building, 1000 Independence Avenue, SW, Washington, DC 20585-1290

Invited presentation on "Introduction and overview of the Basic Energy Science (BES) program."

59b Assuring a Secure Energy Future

Raymond L. Orbach, U. S. Department of Energy, 1000 Independence Ave., S.W, Washington, DC 20585

Invited presentation on "Assuring a secure energy future."

59c Solar Energy Utilization

Nathan S. Lewis, Division of Chemistry and Chemical Engineering, California Institute of Technology, 1200 E. California Blvd., Pasadena, CA 91125

Invited presentation on "Solar energy utilization."

59d Advanced Nuclear Energy Systems

James B. Roberto, Oak Ridge National Laboratory, PO BOX 2008 MS6240, Oak Ridge, TN 37831-6240

Invited presentation on "Advanced nuclear energy systems."

59e Hydrogen Economy

Mildred S. Dresselhaus, Massachusetts Institute of Technology, 77 Massachusetts Ave, Cambridge, MA 02139

Invited presentation on the hydrogen economy.

59f Electrical Energy Storage

Héctor D. Abruña, Department of Chemistry and Chemical Biology, Cornell University,, Baker Lab, Ithaca, NY 14853-1301

Invited presentation on electrical energy storage.

59g Catalysis for Transportation Fuels

Bruce C. Gates, Chemical Engineering and Materials Science, University of California at Davis, One Shield Avenue, Davis, CA 95616

Invited presentation on catalysis for transportation fuels.

Reprinted from CEP (Chemical Engineering Progress), November 2007

Mitigating Global Climate Change Through Power-Generation Technology

Frank Princiotta, U.S. Environmental Protection Agency

The Intergovernmental Panel on Climate Change's (IPCC) Fourth Assessment Synthesis Report, due to be released later this month, is likely to provide a major wake-up call. A summary of the assessment *(1)* published earlier this year concluded, among other things, that there is significant evidence of the inevitable warming of the climate system, and that increases in anthropogenic greenhouse gas (GHG) concentrations are very likely responsible for the observed increase in globally averaged temperature since the mid-20th century.

Several recent events, including the "Climate Change: Impacts, Adaptation and Vulnerability" conference sponsored by the Royal Geographical Society in London and a speech by IPCC chairman Rajendra Pachauri to the United Nations General Assembly *(2)*, provided glimpses into the final synthesis report. That document is expected to paint a picture of urgency: water stresses as a result of the melting of glaciers and major precipitation changes that are already underway will affect millions of people, and a likely unavoidable temperature increase exceeding 1.5–2.5°C will put 20%–30% of existing plant and animal species in danger of extinction.

Various aspects of climate change are addressed extensively in the scientific literature. This article aims to provide a succinct integration of information about the projected warming of the earth in the decades ahead, the emission reductions that may be needed to constrain this warming to tolerable levels, and the potentially available technologies to help achieve these emission reductions. Although it focuses on power generation technologies that can play a significant role in reducing CO_2 emissions, it is important to note that other sectors (*e.g.*, transportation) will need to reduce their emissions as well. Since many of these evolving technologies have higher costs and/or greater operational uncertainties than currently available carbon-intensive technologies, robust policies will need to be in place to encourage their utilization.

Note that unless specifically indicated otherwise, CO_2 concentrations are given in ppmv and "warming" refers to realized or transient warming rather than equilibrium or "eventual warming."

Factors that drive CO_2 emissions

The World Resources Institute (WRI) examined the factors that have driven CO_2 emissions for key countries in the 1992–2002 time period *(3)*. These factors are: gross domestic product (GDP), growth per capita, population growth, carbon intensity growth per unit of energy (more coal in the mix increases this factor), and growth in energy usage per unit of GDP. The sum of these factors approximates the annual CO_2 emissions growth rate.

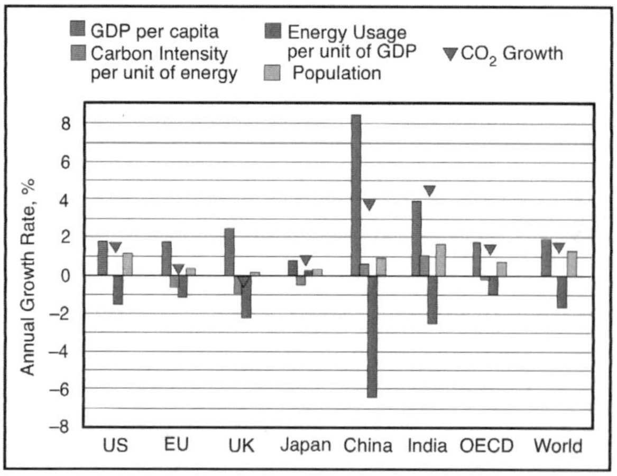

■ Figure 1. Factors driving atmospheric concentrations of CO_2 for selected countries, 1992–2002.

WRI's data were used to generate Figure 1, which shows how these factors have influenced the annual growth rate of CO_2 for selected countries during this ten-year period. Despite decreases in energy use per unit of GDP, the CO_2 growth rate has been about 1.4%/yr worldwide and in the U.S. The growth rates for China and India, however, have both been about 4%/yr; China's increases are driven by economic growth, India's by economic and population growth. In the absence of significant decreases in energy use per unit of economic output, CO_2 emissions growth rates would have been substantially higher.

A recent analysis by Raupach concluded that between 2000 and 2004, worldwide emissions of CO_2 increased at an annual growth rate of 3.2%/yr *(4)*, which is faster than in previous years and faster than had been predicted. Furthermore, it is more than double the growth rate of the 1992–2002 time period. Rapidly developing economies in China and other Asian countries are particularly significant in this recent and troubling trend. China is currently constructing the equivalent of two 500-MW coal-fired power plants per week and a capacity comparable to the entire United Kingdom power grid each year *(5)*.

Figure 2 summarizes these global emission trends.

Projected warming and the uncertainties

In order to estimate warming between now and the year 2100 with any confidence, a credible base case, or business-as-usual scenario, must be established. The analysis presented here is based on the scenario developed by the International Energy Agency (IEA) *(6)*, because it is consistent with current driving forces and does not assume major technology changes over time. It assumes CO_2 emissions increases of 1.6%/yr during the 2000–2030 time interval and 2.2%/yr from 2030 to 2050. To extend the IEA base case to 2100, reduced growth rates of1.2%/yr during the period from 2050 to 2075 and 0.7%/yr for 2075–2100 were assumed, based on projected declines in population growth rates and relatively stable growth rates of GDP, carbon intensity and energy intensity.

Based on this modified emission scenario, Version 4.1 of MAGICC (Model for Assessment of Greenhouse gas Induced Climate Change) *(7)* generated predictions of CO_2 concentrations (Figure 3) and warming (Figure 4) into the year 2100. MAGICC is a set of coupled gas-cycle, climate and ice-melt models that allows the determination of the global mean temperature resulting from user-specified emissions scenarios. Note that the uncertainty ranges in Figures 3 and 4 show that warming forecasts are much more uncertain than concentration-increase forecasts.

When one accounts for the continued warming projected into the next century (Figure 4), the equilibrium (or eventual) warming is expected to range from 3.0 to 8.1°C, with the best guess at 4.8°C, above 1990 levels, assuming that the ultimate steady-state CO_2 concentration will be 1,000 ppm. The main uncertainty factor for warming projections is the extent to which the atmosphere is sensitive to a doubling of CO_2 concentration (*i.e.,* how much does the global equilibrium temperature change if CO_2 concentration doubles?). The extent of warming is quite uncertain, with estimates ranging from 1.5°C to 4.5°C *(1, 7).*

Achievable mitigation levels

Figure 5 presents the April 2007 analysis by IPCC relating projected warming to various global impacts, including fresh water availability, ecosystem damage, food supplies, seawater rise, extreme weather events, and human health impacts *(8)*. Superimposed on the IPCC data are projected warming ranges for a credible business-as-usual case and an aggressive mitigation case. It is significant that current (2007) warming (0.3°C since 1990 and 0.8°C since 1750) has already had measurable impacts.

Base-case warming from 1990 to 2100 is projected to be in the range of 2.2 to 4.7°C, with 3.2°C as the best guess. A temperature increase in this range could potentially result in severe impacts, especially if the temperature increase is in the middle to upper end of the range. Particularly troublesome impacts could include water shortages affecting millions of people, wide-scale ecosystem extinctions, lower food production in many areas, loss of wetlands, damage and mortality from storms and floods, and increased adverse health impacts from infectious diseases.

Given current emission trends, the current unavailability of low-emitting technologies, and the likelihood of inaction in the near term, it is likely that the best result achievable — even with a major global mitigation program (*e.g.,* decreasing CO_2 emissions 1%/yr starting in 2010) — would be to

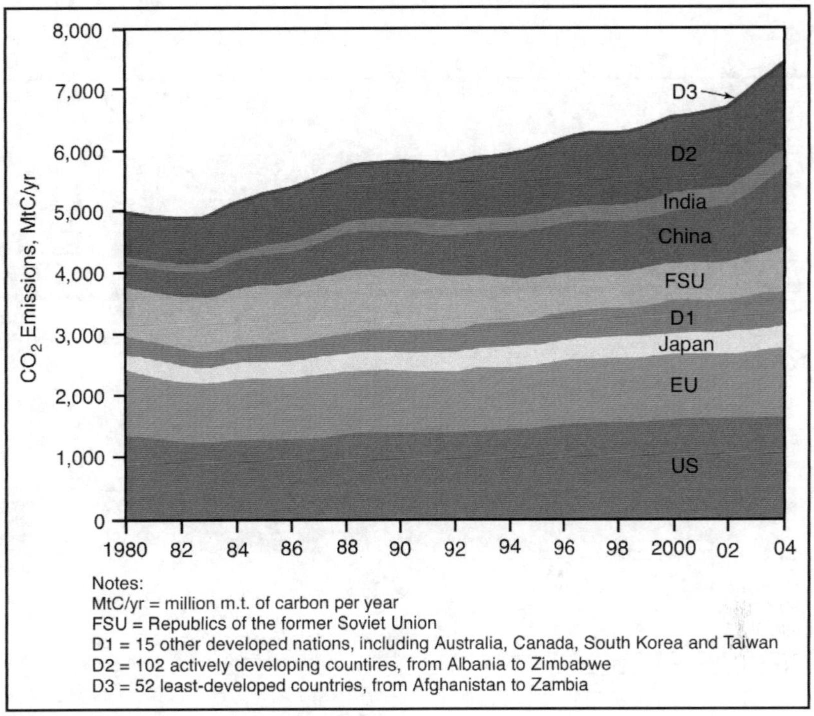

■ Figure 2. CO_2 emissions data by countries.

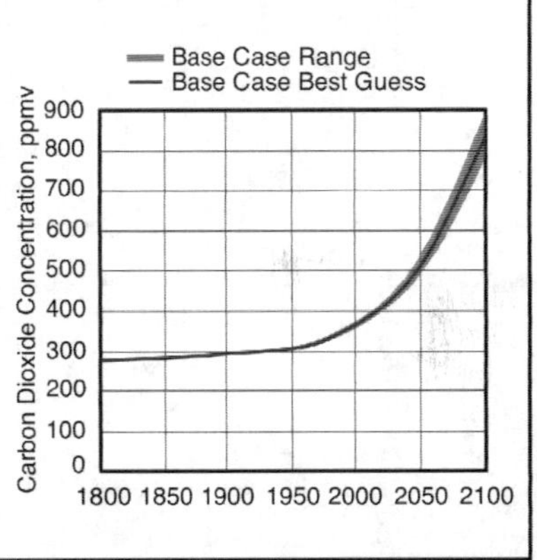

■ Figure 3. Projected CO_2 concentrations for the modified IEA base case.

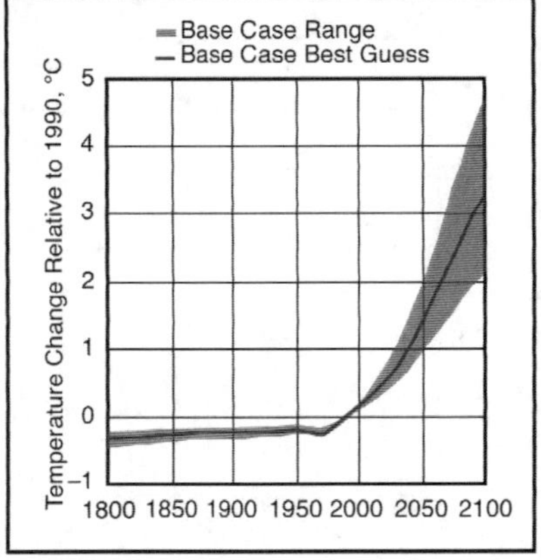

■ Figure 4. Projected warming for the modified IEA base case.

limit warming to about 2°C (or somewhere in the range of 1.3°C to 2.8°C).

Let's consider what reduction levels are required and in what year they would have to start in order to limit warming to less than 2.0°C. Figures 6 and 7, which are based on a large number of MAGICC runs, depict the annual emission reductions required to meet this goal and the corresponding CO_2 concentrations. Note that an annual decrease of 0% means emissions are held constant, or capped — which represents a major mitigation effort relative to the +1.6%/yr increase in CO_2 emissions of the base case.

Major annual reductions in emissions will be necessary to achieve a warming target of 2.0°C or less. The earlier these reductions start, the lower the annual reduction rate can be. For example, if a mitigation program were to start in 2015, reductions would need to be about 1% annually to limit warming to about 2.0°C above the 1990 level, whereas if such a program were to start in 2025, annual reductions of about 1.8%/yr would be needed.

The large uncertainty in the temperature associated with a given maximum CO_2 concentration is illustrated in Figure 8. It shows the range of projected warming from 1990 for a 1%/yr decrease in CO_2 emissions starting in 2010, which is projected to constrain concentrations to the 440–480 ppm range. Note that the high end of this range (2.75°C) is higher than the low end of the business-as-usual case of Figure 4 (approximately 2°C). This highlights the magnitude of the uncertainties in current models.

To illustrate the major challenge associated with such reductions rela-

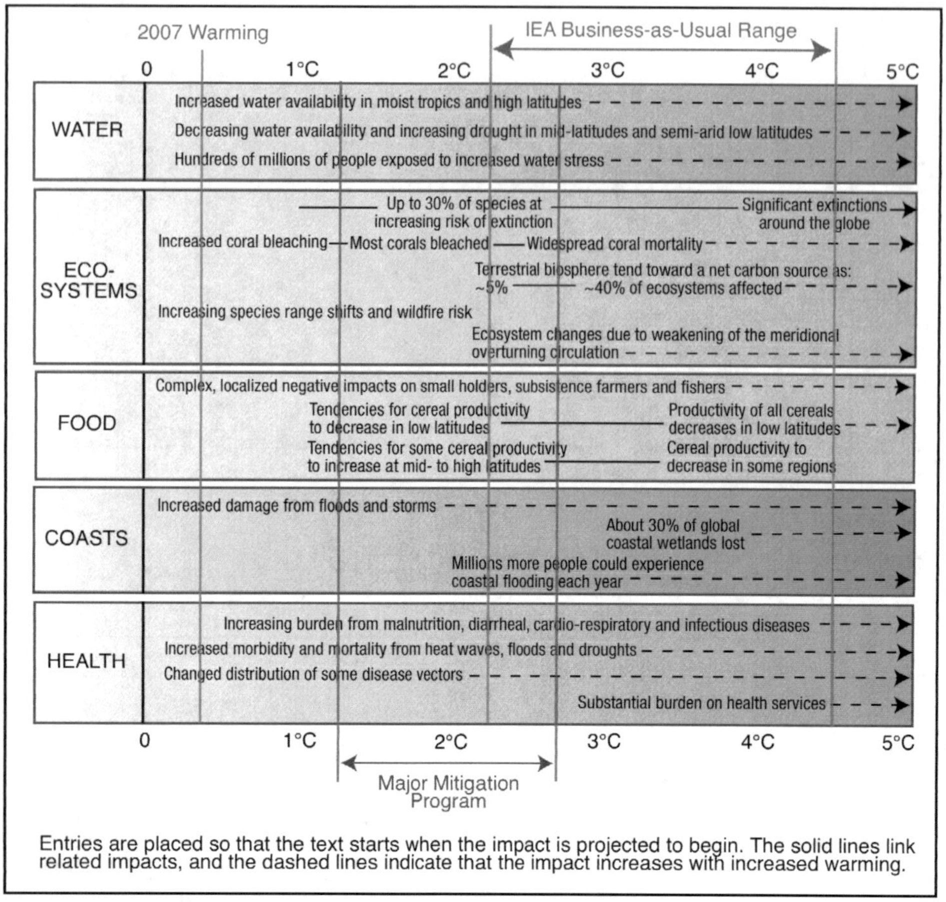

■ Figure 5. Projected impacts as a function of warming from 1990 to 2100.

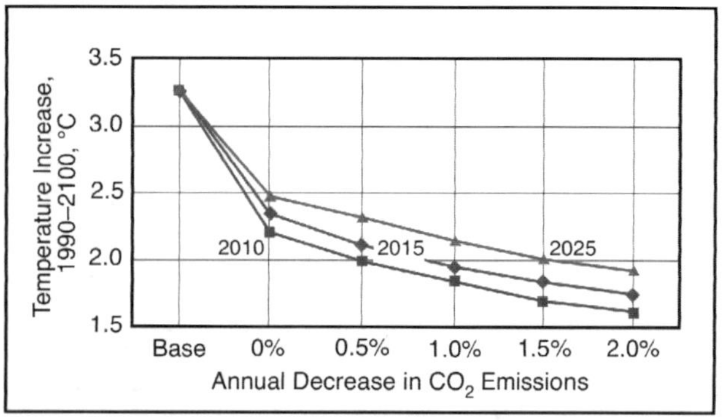

■ Figure 6. Best-guess warming to 2100 as function of annual decrease in CO_2 emissions and year reductions start.

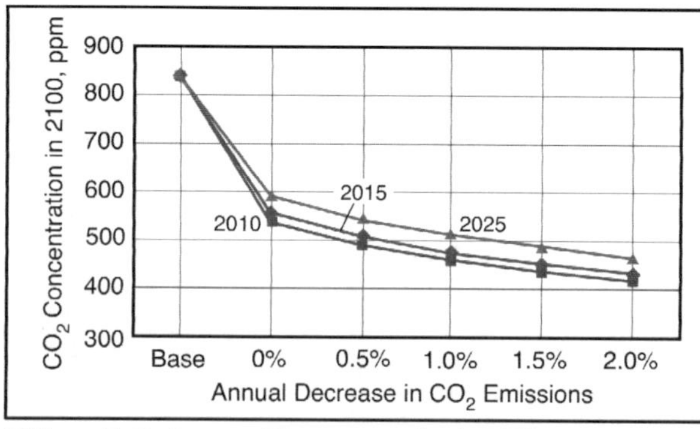

■ Figure 7. Best-guess CO_2 concentration in 2100 as a function of annual decrease in CO_2 emissions and year reductions start.

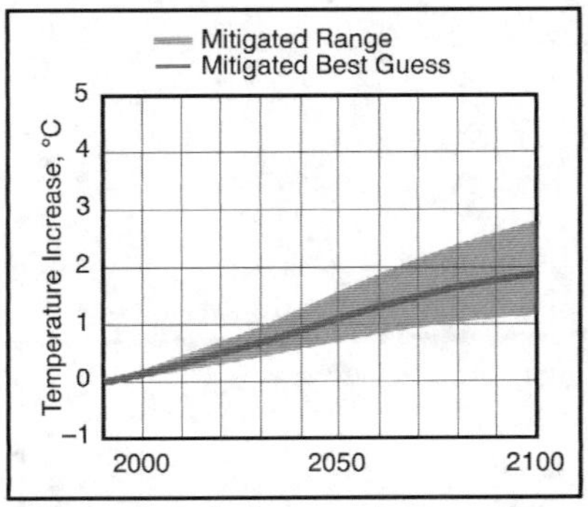

■ Figure 8. Projected warming range for a 1%/yr decrease in CO_2 emissions starting in 2010.

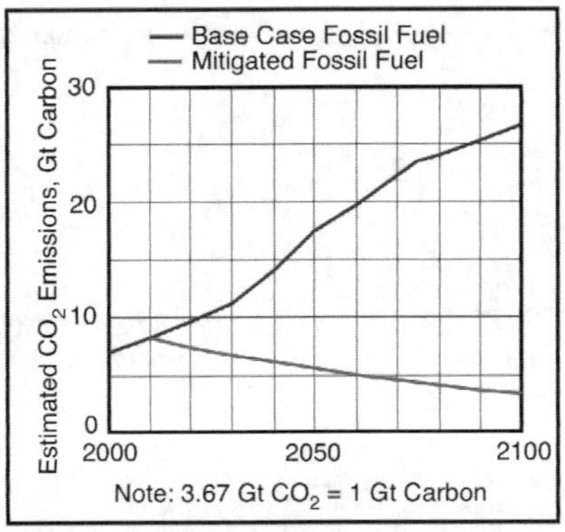

■ Figure 9. Base-case and mitigation scenario to limit CO_2-related temperature increase to <2°C in 2100.

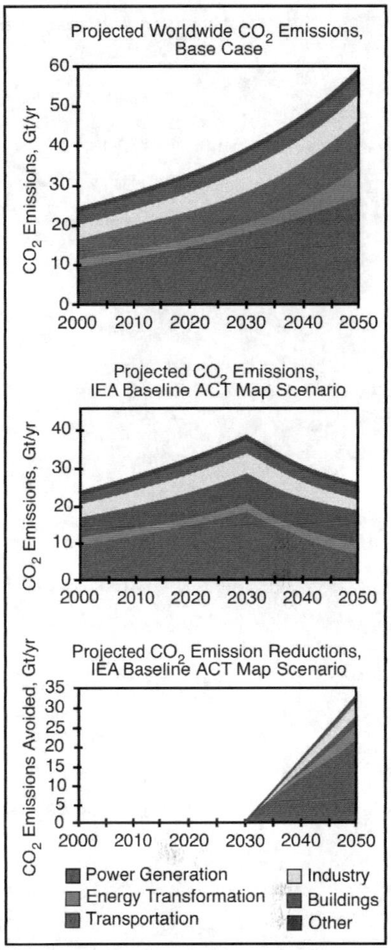

■ Figure 10. IEA projections of CO_2 emissions for the base case and ACT Map control scenario, and the CO_2 emissions avoided.

tive to base-case emission trends, Figure 9 compares the base-case emission trajectory to a mitigation scenario where emissions decrease at a rate of 1%/yr starting in 2010. This would limit CO_2 concentration to about 460 ppm and warming to 1.9°C over 1990 levels. The area between the curves represents the amount of emission reduction needed to achieve the target temperature (relative to the base case), which amounts to more than one trillion tons of carbon or more than 3.7 trillion tons of CO_2 over the 90-year period. This can only be described as a monumental political, social and technological challenge.

The mitigation challenge: Which sectors are most important?

In order to identify the most productive mitigation strategies, one must understand the current and projected sources of CO_2 and other greenhouse gases. Figure 10 presents information derived from IEA *(5)* data.

The top graph in Figure 10 projects world CO_2 emissions by sector for the business-as-usual case. Power generation and transportation sources are expected to be the fastest growing sectors and will be the key to any successful mitigation strategy. There is historical evidence that as a country develops economically, it uses larger quantities of electrical power and experiences a sharp growth in the number and use of motor vehicles and other types of transportation. China and India, with a cumulative population of over 2.4 billion people, are projected to continue their rapid economic expansion with commensurate pressure on the power generation and transportation sectors.

The role of energy technologies

IEA has evaluated what it calls accelerated technology (ACT) scenarios *(6)*. The most optimistic, known as the ACT Map scenario, assumes an aggressive and successful R&D program to develop and improve technologies and a comprehensive technology demonstration and deployment program. It also assumes policies will be in place to encourage the use of these technologies in an accelerated time frame, including various CO_2-reduction incentives to encourage low-carbon technologies with costs up to \$25 per m.t. of CO_2 in all countries from 2030 to 2050. The incentives could take the form of regulation, pricing, tax breaks, voluntary programs, subsidies or trading schemes.

The middle graph of Figure 10 projects CO_2 emissions, and the bottom graph projects CO_2 reductions, for the ACT Map scenario assuming that major technology implementation starts in 2030 when technology is assumed to be deployable. Most of the savings relate to the power generation sector, which includes both production and end-use savings. Note that savings attributed to the transportation sector include the savings associated with transforming less coal, petroleum and natural gas into liquid fuels and the associated CO_2 emissions. This IEA scenario is projected to result in an annual mitigation of 32.5 Gt (1 Gt = 1 billion m.t.) of CO_2 in 2050. As will be discussed later, this level of mitigation would be impossible without the use of improved and, in some cases, breakthrough energy technologies, for both energy production (*i.e.*, power generation) and to enhance end-use efficiency (*e.g.*, lower-emission vehicles).

When the IEA Map scenario is extended to 2100, MAGICC calculations indicate best-guess CO_2 concentrations in 2100 of 500 ppm and corresponding warming of 2.1°C relative to 1990. This is with aggres-

Table. Candidate technologies for CO_2 mitigation during power generation.

Technology	Current State of the Art	2050 Impact, Gt/yr *(6)*	Issues	RD&D Needs	Other Potential Environmental Impacts
Nuclear Power, Next Generation	Developmental, Generation III+ and IV: *e.g.,* pebble-bed modular reactor and supercritical-water-cooled reactor	1.9	Deployment targeted by 2030 with a focus on lower cost, minimal waste, enhanced safety, resistance to proliferation	High Demonstrations of key technologies with complementary research on important issues	Reduction in emissions SO_x, NO_x and fine PM; small but potent and long-lived waste could contaminate a small area
Nuclear Power, Current Generation	Commercial, Generation III: pressurized-water reactors and boiling-water reactors	1.8	Plant siting, high capital costs, levelized cost 10–40% higher than coal or gas plants, potential uranium shortages, safety, waste disposal, proliferation	Medium Waste disposal research	Reduction in emissions of SO_x, NO_x and fine PM; small but potent and long-lived waste could contaminate a small area
Natural Gas Combined Cycle	Commercial, 60% efficiency	1.6	Limited by natural gas availability (which is a major constraint), high efficiency, low capital costs	Medium Higher efficiencies, with new materials desirable	Reduction in emissions of SO_x, NO_x and fine PM; fewer mining impacts and residues for disposal or alternative uses
Wind Power (Renewable)	Commercial	1.3	Costs are very dependent on strength of wind source, large turbines are visually obtrusive, intermittent power source	Medium Higher efficiencies, onshore demonstrations	Reduction in emissions of SO_x, NO_x and fine PM; fewer mining impacts and residues for disposal or alternative uses; extraction R&D could enhance CH_4 availability
Coal IGCC with CCS	IGCC: Early commercialization UGS: Early development	1.3	IGCC: High capital costs, questionable for low-quality coals, complexity, potential reliability concerns; UGS: Cost, safety, efficacy	High IGCC: Demonstrations on a variety of coals , hot-gas cleanup research; UGS: Long-term demonstrations to evaluate many geological formations for efficacy, cost and safety	Lower power plant efficiency yields higher emissions of SO_x, NO_x and fine PM and higher coal mining impacts
Pulverized Coal/Oxygen Combustion with CCS	Developmental	1.3	Oxygen combustion allows lower-cost CO_2 scrubbing but oxygen production cost is high; UGS: Cost, safety and permanency	High Large-scale pilot tests followed by full-scale demonstrations, low-cost oxygen production; UGS: Long-term demonstrations to evaluate many geological formations for efficacy, cost and safety	Lower power plant efficiency yields higher emissions of SO_x, NO_x and fine PM and higher coal mining impacts

IGCC = Integrated Gasification Combined Cycle
CCS = CO_2 Capture and Storage
UGS = Underground Geological Storage
PM = Particulate Matter

sive research, development, demonstration and deployment programs as assumed by IEA as well as the optimistic assumption of further major emission reductions (2%/yr) for 50 years beyond the IEA time frame of 2050. This suggests that even a major global mitigation program based on successful development and deployment of several new technologies will still allow substantial global warming by 2100.

Let us now look at the critical power sector and examine the technology options available, their current state of development, and the R&D required for them to meet their potential to reduce CO_2 emissions.

Focusing on power generation

The power generation sector, which is projected to grow at a rate of 2%/yr, provides the greatest opportunity for reduction of CO_2 emissions in the coming decades. However, there are major capacity expansions underway for coal-fired power generation

Table. Candidate technologies for CO_2 mitigation during power generation — continued.

Technology	Current State of the Art	2050 Impact, Gt/yr *(6)*	Issues	RD&D Needs	Other Potential Environmental Impacts
Pulverized Coal/Oxygen Combustion with CCS	Developmental	1.3	Oxygen combustion facilitates carbon capture, but oxygen production cost is high; UGS: Cost, safety and permanency	High Large-scale pilot tests followed by full-scale demonstrations, low-cost oxygen production; UGS: Long-term demonstrations to evaluate many geological formations for efficacy, cost and safety	Lower power plant efficiency yields higher emissions of SO_x, NO_x and fine PM and higher coal mining impacts
Pulverized Coal with CCS	UGS: Developmental CO_2 Scrubbing with MEA: Near commercial but too expensive	1.3	UGS: Cost, safety, efficacy, CO_2 scrubbing is energy-intensive so costs are unacceptably high	High Affordable CO_2-removal technologies need to be developed and demonstrated UGS: Long-term demonstrations to evaluate many geological formations for efficacy, cost and safety	Lower power plant efficiency yields higher emissions of SO_x, NO_x and fine PM and higher coal mining impacts
Solar Photovoltaic and Concentrating (Renewable)	First-generation commercial, but very high costs	0.5	Costs are unacceptably high, solar resource is intermittent in many locations	High Breakthrough RD&D needed to obtain cells with higher efficiency and lower capital costs	Reduction in emissions of SO_x, NO_x and fine PM; fewer mining impacts and residues for disposal or alternative uses
Biomass as Fuel and Co-fired with Coal (Renewable)	Commercial, steam cycles	0.5	Biomass is a dispersed source, limited to 20% when co-fired with coal	Medium Biomass/IGCC would enhance efficiency and CO_2 benefit, genetic engineering to enhance biomass plantations	Reduction in emissions of SO_x, NO_x and fine PM; fewer mining impacts and residues for disposal or alternative uses; potential ecological impacts from biomass plantations
Hydroelectric (Renewable)	Commercial	0.5	High capital costs, potential ecological disruption, siting challenges	Medium Minimize environmental footprint	Ecosystem impacts
More-Efficient Coal-Fired Power Plants	Early commercialization of supercritical and ultra-supercritical	0.2	Currently maximum efficiency of 45%, yielding 36% less CO_2 than current fleet	High New affordable materials are needed to enhance efficiency to 50–55%	Small reduction in emissions of SO_x, NO_x and fine PM; fewer mining impacts and residues for disposal or alternative uses
Coal IGCC with no CCS	IGCC: Early commercialization	0.2	IGCC: High capital costs, complexity and reliability concerns, only modest CO_2 reductions without CCS	High Demonstrations on a variety of coals, hot-gas cleanup research	Small reduction in emissions of SO_x, NO_x and fine PM; fewer mining impacts and residues for disposal or alternative uses

in China, India and other countries. Since those plants do not have CO_2-mitigation technology planned and can have lifetimes up to 50 years, the sooner technology is ready for implementation and mandated, the sooner new plants can incorporate that technology and control emissions. Current retrofit technology is theoretically available, but it will likely be substantially more expensive per unit of power generated than the cost to outfit new plants with CO_2 capture or advanced CO_2-removal technologies, which are now in the early development stages.

Major reductions in CO_2 concentrations can result from lower emissions during power generation and from lower electricity usage due to improved end-use efficiency. The table summarizes major power generation options that offer significant opportunities for CO_2 mitigation. The options are presented in the order of highest potential for CO_2 mitigation consistent with the IEA's ACT Map scenario.

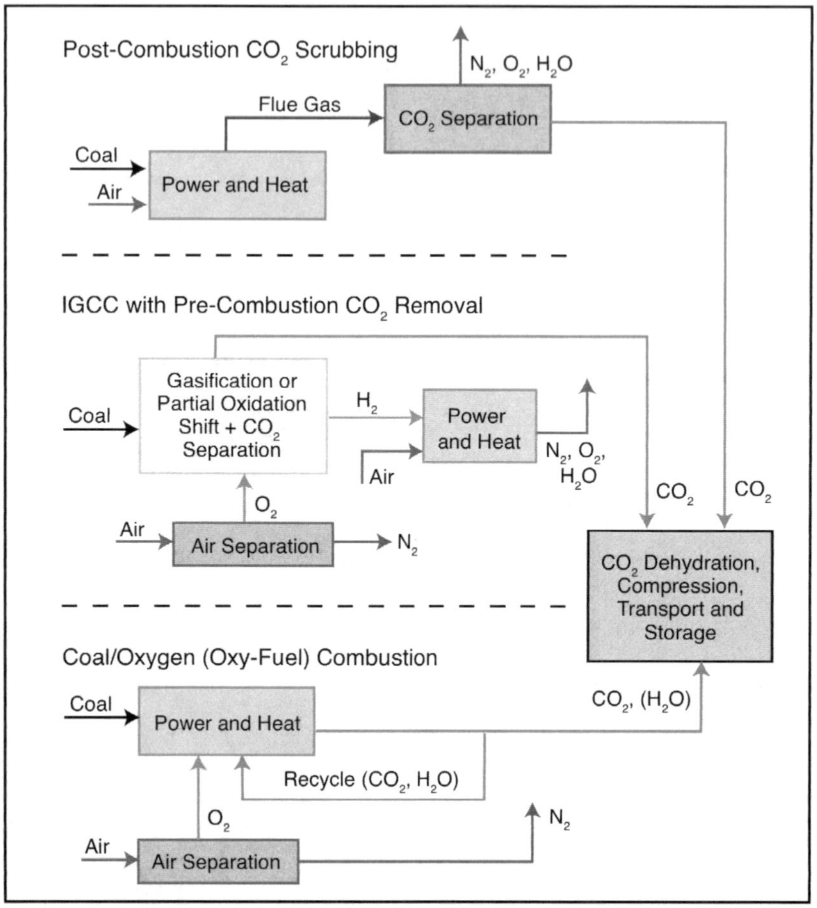

■ Figure 11. Three key technologies for capturing CO_2 from coal-fired power plants.

The table includes the IEA's projected CO_2 reductions for each technology in Gt/yr of CO_2 in 2050. To put these numbers in perspective, to achieve the mitigation depicted in Figure 9 by 2050, total reductions of about 40 Gt of CO_2 (10.9 Gt carbon) are required, and for the less-aggressive IEA Map scenario, 32 Gt of CO_2 mitigation would be needed.

Key generation technologies include nuclear power, natural gas combined cycle, and three coal-combustion technologies (integrated gasification combined cycle [IGCC], pulverized coal/oxygen combustion, and conventional pulverized coal), all with integrated CO_2 capture and underground storage. The three coal-based technologies (Figure 11) are quite important — the IEA scenario projects each of the three to reduce emissions by 1.3 Gt/yr of CO_2 in 2050.

IGCC technology is the primary focus of the U.S. R&D program, but it requires complex chemical processing and pure oxygen for the gasification process, and it cannot be readily retrofitted to existing plants. Oxy-combustion systems also require pure oxygen for combustion, but they are less complex and have the potential for retrofitting at existing plants. CO_2 removal by scrubbing, adsorption or membrane separation is conceptually simple and is inherently retrofitable, but these technologies are in the early stages of development, and commercial amine scrubbers use large quantities of energy for sorbent regeneration and are expensive.

The Massachusetts Institute of Technology (MIT) recently completed an in-depth study of coal in a carbon-constrained world and concluded that CO_2 capture and sequestration is the critical enabling technology that would reduce CO_2 emissions significantly while also allowing coal to meet the world's pressing energy needs *(5)*. The MIT researchers also concluded that funding for research to demonstrate an integrated system of capture, transportation and storage of CO_2 at full scale is currently inadequate.

With the exception of wind power, renewable technologies are not expected to have major mitigation impacts in the 2050 time frame. Solar generation is projected to be prohibitively expensive unless there is a major research breakthrough. Major utilization of biomass is likely to be limited by the technology's dispersed nature, its low energy density, and competition for the limited resource in the transportation sector.

The R&D needs of the power generation sector are critical, especially in the areas of CO_2 capture and storage (CCS) and the next generation of nuclear power plants. All three of the capture technologies described above warrant aggressive R&D programs. Because of the many uncertainties associated with IGCC, R&D efforts should not focus on that technology alone. Much more emphasis should be placed on pulverized coal/oxygen (oxy-fuel) combustion and high-efficiency pulverized coal with CO_2 flue-gas capture technology.

Underground sequestration (which is needed for each of these technologies) has extraordinary potential. However, it is currently in an early developmental stage, and there are a host of questions that can only be resolved through a major demonstration program focusing on the geological formations with the greatest potential capacity. MIT estimates that three full-scale CCS projects in the U.S. and ten worldwide are needed to cover the range of likely accessible geologies for large-scale storage *(5)*.

Advanced nuclear power technology is quite promising and could start making a major impact by 2030.

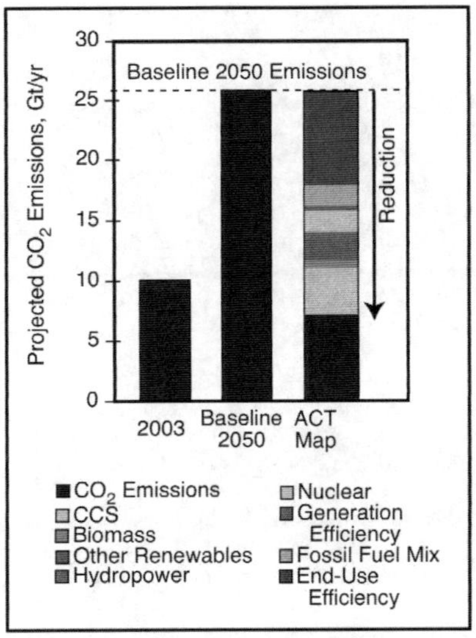

■ Figure 12. Projected CO_2 emission reductions in the power-generation sector by various technologies under the ACT Map scenario.

However, several successful demonstrations are needed to resolve remaining technical problems, instill confidence in the utility industry that the technology is affordable and reliable, and assure the public that it is safe.

Figure 12 summarizes IEA's projections of the impacts of key technologies on CO_2 emissions from the power generation sector in 2050. Assuming aggressive R&D and incentive programs, it is expected that end-use efficiency improvements, carbon capture and sequestration, and nuclear power will play important roles in that time frame.

Adequacy of R&D

Various authors *(5, 6, 9, 10, 11)* have observed that R&D funding in the energy area needs to be substantially increased in order for key CO_2-emission-reduction technologies to be commercially ready in the time frame needed to limit atmospheric concentrations to protective levels. As noted earlier, the later a mitigation program is initiated, the more severe emission cuts will need to be if CO_2 concentrations above 450–500 ppm are to be avoided. Recently, the Stern Report concluded that support for energy R&D should at least double, and support for the deployment of new low-carbon technologies should increase up to five-fold *(12)*.

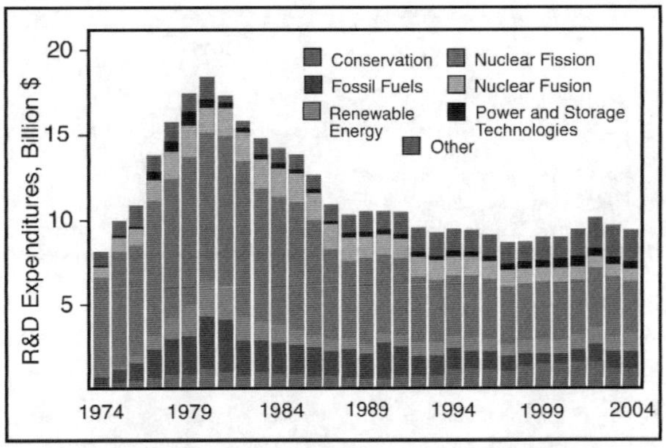

■ Figure 13. World R&D expenditures for key energy sectors (2004 dollars and exchange rates).

Figure 13, generated from IEA data *(13)*, depicts world research expenditures in critical energy technology. Global R&D spending decreased substantially relative to the 1976–1987 period and has been fairly flat in recent years. That 12-year funding spike was influenced by the Arab oil embargo that started in October 1973.

In the last few years, the U.S. has redirected some of its limited research resources to several key technologies, especially hydrogen/fuel cells, IGCC, CCS and, most recently, biomass-to-ethanol processes. The U.S. has coordinated its efforts in these areas through the Climate Change Technology Program (CCTP; www.climatetechnology.gov). Within the constraints of current budget priorities, the CCTP has coordinated a diversified portfolio of advanced technology R&D focusing on: energy efficiency enhancements; low-GHG-emitting energy supply technologies; CCS and carbon sequestration methods; and technologies to reduce emissions of non-CO_2 gases. Also, the U.S. EPA *(14)* is implementing a series of voluntary programs that encourage the reduction of GHG emissions, including Energy Star for the building sector, transportation programs, and non-CO_2 emission-reduction programs in collaboration with industry. These efforts could provide a foundation for an expanded program consistent with the overall mitigation challenge.

Most of the non-coal technologies offer the potential for lower air emissions, water effluents and waste residues. IGCC also offers the potential for such benefits; however, incorporating CO_2 capture, transport and storage substantially decreases overall plant efficiency, with the potential for commensurate increases in coal-related pollution per unit of power generated. The transportation technologies all offer the potential for reducing U.S. dependence on foreign oil.

The countries that can bring these technologies to market first have the potential for major revenue streams from a huge international market.

Article continues with Literature Cited on next page

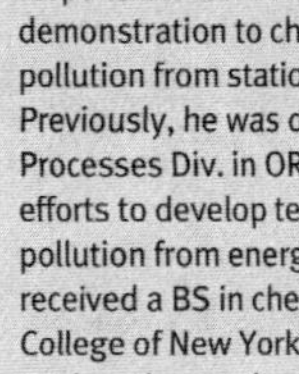

FRANK PRINCIOTTA is director of the Air Pollution Prevention and Control Div. in the Office of Research and Development (ORD) at the U.S. Environmental Protection Agency (109 T. W. Alexander Dr., Research Triangle Park, NC 27711; Phone: (919) 541-2821; Fax: (919) 541-5227; E-mail: princiotta.frank@epa.gov). He is responsible for research, development and demonstration to characterize and control air pollution from stationary and mobile sources. Previously, he was director of the Energy Processes Div. in ORD, where he managed EPA efforts to develop technology to control pollution from energy production. Princiotta received a BS in chemical engineering from City College of New York and a certificate in nuclear engineering graduate studies from the Oak Ridge School of Reactor Technology. He is chair of the Triangle Area Research Directors Council and a member of the Air and Waste Management Association. EPA has awarded him bronze, regulatory support and gold medals, and the President has conferred on him the Meritorious Executive Award on two occasions. He is the author of articles on air pollution technology, a frequent speaker before professional conferences, and has testified before Congress.

Literature Cited

1. **Intergovernmental Panel on Climate Change,** "Climate Change 2007: The Physical Science Basis, Summary for Policymakers, Fourth Assessment Report of the Intergovernmental Panel on Climate Change," IPCC, Geneva, Switzerland, available at www.ipcc.ch/SPM2feb07.pdf (2007).
2. **Pachauri, R.,** speech presented at "The Future in Our Hands: Addressing the Leadership Challenge of Climate Change," New York, NY, available at www.un.org/webcast/climatechange/high-level/ or at www.ipcc.ch/ (Sept. 24, 2007).
3. **World Resources Institute,** "Climate Analysis Indicators Tool (CAIT)" Version 3.0., WRI, Washington, DC, available at http://cait.wri.org (2006).
4. **Raupach, M. R., *et al.*,** "Global and Regional Drivers of Accelerating CO_2 Emissions," Proceedings of the National Academy of Sciences of the United States of America, available at www.pnas.orgcgidoi10.1073pnas.0700609104 (May 22, 2007).
5. **Massachusetts Institute of Technology,** "The Future of Coal, An Interdisciplinary MIT Study," MIT, Cambridge, MA (2007).
6. **International Energy Agency,** "Energy Technology Perspectives 2006," Organization for Economic Cooperation and Development, IEA, Paris, France (2006).
7. **Wigley, T. M. L, and S. C. B. Raper,** "Interpretation of High Projections for Global-Mean Warming," *Science,* **293,** pp. 451–454; MAGICC (Model for the Assessment of Greenhouse-gas Induced Climate Change) can be downloaded from www.cgd.ucar.edu/cas/wigley/magicc/ (2001).
8. **Intergovernmental Panel on Climate Change,** "Climate Change 2007: Impacts, Adaptation and Vulnerability, Summary for Policymakers, Fourth Assessment Report of the Intergovernmental Panel on Climate Change," IPCC, Geneva, Switzerland, available at www.ipcc.ch/SPM13apr07.pdf (2007).
9. **Hawksworth, J.,** "The World in 2050; Implications of Global Growth for Carbon Emissions and Climate Change Policy," PriceWaterhouseCoopers, London, U.K. (Sept. 2006).
10. **Morgan, G., *et al.*,** "The U.S, Electric Power Sector and Climate Change Mitigation," The Pew Center on Global Climate Change, Arlington, VA (June 2005).
11. **Princiotta, F. T.,** "Renewable Technologies and Their Role in Mitigating Greenhouse Gas Warming," U.S.-Dutch Symposium: Facing the Air Pollution Agenda for the 21st Century," Vol. 72, Elsevier Science Publishers, Amsterdam, The Netherlands (1998).
12. **Stern, N.,** "Stern Review on The Economics of Climate Change," 2006 Pre-publication Version, HM Treasury, London, U.K., available at www.hm-treasury.gov.uk/independent_reviews/stern_review_economics_climate_change/stern_review_report.cfm (2006).
13. **International Energy Agency,** "RD&D Budgets," available at www.iea.org/RDD/TableViewer/tableView.aspx?ReportId=1
14. **U.S. Environmental Protection Agency,** "Current and Near-Term Greenhouse Gas Reduction Initiative," U.S. EPA, Washington, DC, www.epa.gov/climatechange/policy/neartermghgreduction.html.

CEP

68b Vacuum Swing Adsorption Cycles for Carbon Dioxide Capture from Flue and Stack Gases

James A. Ritter, Amal Mehrotra, and Armin D. Ebner. Department of Chemical Engineering, University of South Carolina, Columbia, SC 29208

There are many misconceptions about and misleading reports on the viability of pressure swing adsorption (PSA) cycles for CO2 capture from flue gas. These misconceptions and misleading reports stem largely from a 1994 International Energy Agency report [1] that evaluated CO2 separation and capture from flue gas using 13X zeolite in both pressure swing adsorption (PSA) and temperature swing adsorption (TSA) processes. They concluded that PSA and TSA are too energy intensive for use with gas and coal fired power systems. This conclusion has led others to extrapolate their findings and further conclude that adsorption systems, in general, are not applicable for CO2 separation and capture from flue gas. This disturbing misinformation has perpetuated for nearly a decade with essentially the same misleading statements being made in a 2003 IEA report [2]. However, it is strongly contended that those reports might have been factually inaccurate, that some grave errors might have been made in the evaluation of cyclic adsorption technologies for CO2 capture from flue gas, and that PSA might be very applicable to the capture of CO2 from flue gas. These claims are substantiated with evidence from works found in the literature and from some of the author's recent works.

First and foremost, a recent report by Izumi [3] from Mitsubishi Heavy Industry indicated that CO2 was being recovered from flue gas commercially using a layered PSA bed containing X and A type zeolites and activated carbon. Although, not much detail about the PSA cycle was provided, this news was exciting. The Japanese must have discovered something positive about using PSA for CO2 capture from flue gas that has been overlooked by investigators in the USA and apparently everyone else around the world!

Second, in a series of works by the Ritter and co-workers [4-7], novel PSA cycles have been explored systematically for the capture of CO2 from flue gas at high temperature using a K-promoted hydrotalcite like compound (HTlc) as the adsorbent. They revealed that a variety of vacuum swing adsorption (VSA) cycles could be used to concentrate CO2 to over 90 vol% at over 90% recovery from a feed stream containing 15 vol% CO2. The most effective cycles were based on the heavy reflux concept, a concept that has been largely missing from the PSA literature. What is really exciting about these high temperature heavy reflux PSA cycles is that they are completely applicable to ambient temperature adsorbents like zeolites and activated carbons. Moreover, the Ritter and co-workers are still working on the problem and have discovered some better heavy reflux PSA cycles for CO2 capture from flue gas that will be discussed during this presentation. They also feel that the lack of information on and therefore the lack of understanding of the heavy reflux PSA concept for concentrating the heavy component from a feed stream might be the principle reason why the IEA reports were mistaken in their assessments of PSA for CO2 capture from flue gas. To this end, the heavy reflux concept will also be discussed during this presentation

Third, some very recent work by Webley and co-workers [8] substantiated the claims made by the Ritter and co-workers over the past few years on the economic viability of PSA cycles for CO2 capture from flue gas, especially VSA cycles. They proclaimed that the information about the cost of PSA provided in the 2003 IEA report [2] was in error. Zhang et al [8] recalculated the cost of PSA to be around $67/tonne of CO2 produced compared to that in the IEA report of $97/tonne. This new cost for PSA, as verified through the IEA, compared much more favorably with amine scrubbing technology at $60/tonne. What is most exciting about this recent set of events is that the PSA cycles analyzed by the IEA were not designed for heavy component recovery, like CO2; instead, they were designed for light component recovery, like CH4 from a CH4/CO2 mixture.

As alluded to above, there are significant differences between these two kinds of PSA cycles, i.e., between the ones designed for light component enrichment and recovery and those designed for heavy component enrichment and recovery, with understanding of the later lacking, even by some of the best PSA researchers in the field. To make matters worse, the design of a multi-bed PSA process with complex cycle sequencing, like that envisioned for CO2 capture from flue gas, is a non-trivial exercise. PSA beds are always coupled together, usually contain more than one layer of adsorbent, and operate sequentially with each undergoing cycle steps such as pressurization, feed, heavy reflux, equalization, depressurization, light reflux, and repressurization. As an added complication, each cycle step is not necessarily of the same duration. These features make the number of possible cycle configurations very large; and unfortunately, design strategies on how to best configure such a complex PSA cycle is more of an art than a science. To this end, this presentation will also introduce some complex VSA cycle sequences that might be applicable for CO2 capture from flue and stack gases.

References:

1. International Energy Agency, “Carbon Dioxide Capture from Power Stations,” 1994.

2. International Energy Agency, “Carbon Dioxide Capture at Power Stations and Other Major Point Sources,” 2003.

3. J. Izumi, “Adsorption technology applications to the industries in Japan,” Abstract and presentation at the 8th International Conference on Fundamentals of Adsorption, Sedona, AR, May (2004).

4. S. P. Reynolds, A. D. Ebner and J. A. Ritter, New Pressure Swing Adsorption Cycles for Carbon Dioxide Sequestration,” Adsorption, 11, 531-536 (2005).

5. S. P. Reynolds, A. D. Ebner and J. A. Ritter, “Carbon Dioxide Capture from Flue Gas by PSA at High Temperature using a K-Promoted HTlc: Effects of Mass Transfer on the Process Performance,” Environmental Progress, 25, 334-342 (2006).

6. S. P. Reynolds, A. D. Ebner, and J. A. Ritter, “Stripping PSA Cycles for CO2 Recovery from Flue

Gas at High Temperature Using a Hydrotalcite-Like Adsorbent," Ind. Eng. Chem. Res., 45, 4278-4294 (2006).

7. S. P. Reynolds, A. Mehrotra, A. D. Ebner and J. A. Ritter, "Heavy Reflux PSA Cycles for CO2 Recovery from Flue Gas. Part I. Performance Evaluation," Adsorption, submitted (2007).

8. J. Zhang, P. A. Webley, and P. Xiao, "Effect of Process Parameters on Power Requirements of Vacuum Swing Adsorption Technology for CO2 Capture from Flue Gas," Energy Conversation and Management, in press on line (2007).

68c Metal Organic Frameworks for Carbon Dioxide Capture

Annabelle I. Benin[1], John J. Low[2], Syed Faheem[1], Antek G. Wong-Foy[3], Kyoungmoo Koh[3], Adam J. Matzger[3], A. Ozgur Yazaydin[4], Randall Q. Snurr[4], Stefano Brandani[5], Jian Liu[6], M. Douglas LeVan[6], and **Richard R. Willis**[1]. (1) New Materials Research, UOP LLC, a Honeywell Company, 50 East Algonquin Road, Des Plaines, IL 60017, (2) Advanced Characterization, UOP LLC, a Honeywell Company, 50 East Algonquin Road, Des Plaines, IL 60017, (3) Department of Chemistry, University of Michigan, 930 N. University, Ann Arbor, MI 48109, (4) Department of Chemical & Biological Engineering, Northwestern University, 2145 Sheridan Road, Evanston, IL 60208, (5) School of Engineering and Electronics, University of Edinburgh, Sanderson Building, The King's Buildings, EH9 3JL, Edinburgh, United Kingdom, (6) Department of Chemical Engineering, Vanderbilt University, VU Station B 351604, 2301 Vanderbilt Place, Nashville, TN 37235-1604

Our goal is to incorporate extremely high pore volume MOF (metal organic framework) materials into a commercially viable carbon dioxide capture from coal-fired power plant flue gas process. Owing to their very open, accessible and tunable porosity and favorable heats of adsorption, several MOFs show promise in this application. Our experimental approach involves a combination of molecular modeling with synthesis of novel MOFs. This talk will focus on the synthesis, characterization, and stability and adsorption application testing of MOFs for this application.

MOFs are a relatively new class of nanoporous materials synthesized in a rational, "building-block" approach by self-assembly of metal or metal oxide vertices interconnected by rigid organic linker molecules. This rational synthesis approach has opened up the possibility of incorporating a wide variety of metal and metal oxide vertices, as well as various functional groups into a MOF structure. Significant progress toward a fundamental understanding of the characteristics required for a MOF suitable for flue gas operation has been achieved. For example, it is now clear that ultra-high surface area MOFs are not suitable for ambient temperature and sub-atmospheric pressure conditions typical for flue gas. A molecular modeling approach on MOF hydrolysis that correlates with experimental data will be provided. The experimental data were generated by subjecting MOF samples to relevant steaming conditions in a combinatorial chemistry heat treatment unit followed by analysis in a high-throughput X-ray diffractometer. Additional results were collected via an in situ high temperature, steam-atmosphere equipped X-ray diffractometer. Many types of MOF materials were prepared and evaluated, ranging from compounds from the open literature to new families of MOF materials prepared via combinatorial and other in-house synthesis strategies. We also initiated the development of a Raman spectroscopic-based method of determining binary adsorption isotherms. Finally, detailed molecular modeling of adsorption isotherms and MOF hydrolysis show that theory and experiment match closely.

68d Economic CO2 Capture with Solid Sorbents

James C. Fisher II and Steven S.C. Chaung. Chemical Engineering, University of Akron, 200 E. Buchtel Commons, Akron, OH 44256-3906

Abstract:

The adsorption and desorption of CO_2 on amine-grafted sorbent has been studied by in situ infrared spectroscopy coupled with mass spectrometry. Infrared spectroscopy demonstrated CO_2 adsorption on amine grafted sorbent as bidentate carbonate, bidentate and monodentate at 50 °C and released at 135 °C. Mass spectroscopy revealed an adsorption capacity of 824μmol CO_2 / g-sorbent. The solid sorbent developed in this paper is intended as a downstream component in a stationary power plant which uses low pressure 135 °C steam to regenerate.

Introduction:

The available approaches for CO_2 capture and separation include absorption of CO_2 in aqueous amines, membrane separation, and adsorption on solid sorbents. The current aqueous amine and membrane technologies are cost-effective for separation of CO_2 from natural gas in liquefaction process and ammonia synthesis process due to the high value of end products. These current technologies when applied for CO_2 capture from coal-fired power plants, increases the cost of electricity by more than 70%. The cost of CO_2 sequestration can be reduced if an effective CO_2 capture sorbent is developed which has (i) high CO_2 adsorption capacity (> 1000 μmole/g), (ii) long term regeneration capacity in power plant flue gas environment, and (iii) low energy requirement for regeneration compared to large amount of energy required for aqueous amine process.

Our previous study has shown that the amine-grafted SBA-15 has one of the highest CO_2 adsorption capacities compared to the solid sorbents reported in literature. The key issues that need to be addressed for development of an effective solid amine sorbent for CO_2 capture include:

- effect of the nature of amine and adsorbed CO_2 species on the CO_2 adsorption capacity of the sorbent
- interaction of the support and amine
- thermal stability and long term CO_2 adsorption/regeneration capacity.

The objectives of this paper are (i) to investigate the nature of adsorbed CO_2 species on the amine-grafted sorbent and a commercial polymer amine based sorbent and to determine the thermal stability of the sorbent. The CO_2 adsorption/desorption is studied by transient technique and temperature-programmed desorption with diffuse reflectance infrared Fourier transform spectroscopy (DRIFTS) coupled with mass spectrometry (MS). The thermal stability of the sorbents is studied by temperature-programmed degradation with DRIFTS and MS.

Exponential methods:

Preparation of amine impregnated sorbent: Beta zeolite was prepared by heating to 150°C for 8 hours to remove adsorbed water. The cleaned zeolite was submersed in 25 mL of tetraethylenepentamine at 150 °C and stirred for 24 hours in an inert environment. Vacuum filtration and 3 10mL ethanol washings were required to remove the excise amine. The powder was the grinded and passed through a 100 mesh sieve.

Adsorption: The experimental setup shown in Fig. 1 consists of a gas manifold which includes a 4-port valve and mass flow controllers, a DRIFTS reactor filled with 30 mg of sorbent, and a mass spectrometer for effluent analysis. The adsorption of CO_2 was carried out by switching the inlet flow from an inert gas stream (Ar) flowing at 30 cc/min to the adsorbing gas stream (10% CO_2 in air) using a 4 port valve. The 4-port valve allows a smooth switch from one flow stream to another. The inert gas stream was bubbled through a H_2O saturator at room temperature corresponding to 4% H_2O in the gas streams during adsorption and desorption. Upon saturation of the sorbent with adsorbing gas, the inlet stream was switched back to the inert gas stream. The adsorption process was preformed twice non-hydrated, twice with H_2O and twice with D_2O with regeneration occurring between each adsorption.

Temperature-programmed desorption: The sorbent containing adsorbed CO_2 was regenerated via temperature-programmed desorption (TPD) by heating the DRIFTS from 30 °C to 135 °C in Ar stream at a rate of 10 °C/min. The CO_2 concentration profile during TPD was monitored by the MS and the amount of CO_2 was quantified by calibrating the CO_2 (m/e=44) response on the MS. The calibration factor was obtained by injecting 1 mL of CO_2 gas in flowing Ar stream using the 6-port valve and calculating the area corresponding to the amount of CO_2 injected. IR spectra have been reported for every 20 °C while heating. The adsorption and TPD were cycled multiple times to check for degradation.

Benzene surface investigation: The amine impregnated zeolite that was used for the capture experiments and a fresh zeolite without amines that was heated to 135 C in Ar where used for benzene surface probing. Benzene adsorption was preformed via Ar flow at 30 cc/min through a benzene saturator for 2 minutes followed by a step switch to pure Ar. The IR bench was used in rapid scan mode to collect spectra 1 minute before, during the benzene introduction and for 17 minutes after gaseous benzene as removed.

Results and discussion:

Pretreatment: A summary of band assignments related to the amine, adsorbed CO2 species and benzene are presented in Table 1. Monodentate bicarbonate at 1407 and 1488 cm-1 with bidentate carbonate at 1564 cm-1 peak are the most evident during the pretreatment which indicates they might be the most preferred adsorbed species under atmospheric conditions. The desorption of CO2 during the heating caused negative bands in the IR difference spectra between wavenumbers 1300 to 1575 cm-1 suggesting the removal of carbonate species and the previously carbonate-covered amine appears between 2800 to 3400 cm-1 with positive bands.

Temperature Programmed Desorption (TPD): Fig. 2 depicts the difference spectra of the non-hydrated adsorbed CO2 as carbonate. CO2 adsorbed as the species bidentate carbonate at 1564 cm-1, monodentate bicarbonate at 1488 and 1407 cm-1 and monodentate carbonate at 1321 cm-1. Monodentate carbonate peak appeared at 120 C and bidentate carbonate and monodentate bicarbonate appeared at 100 C suggesting the later carbonate has a weaker binding. The MS profile for CO2 showed a shoulder at the low temperature side suggesting some form(s) of carbonates may desorbed at different temperatures which is in agreement with the IR spectra. The quantified amount of CO2 adsorption from this experiment was 890 μmol/g-sorbent.

The bidentate carbonate band intensity was stronger and the monodentate bicarbonate and monodentate carbonate appear weaker when the sample was hydrated. The hydreated sample lacked the low temperature shoulder on the MS profile combined with the IR suggests a different preference in adsorption with and without hydration. However, the effect of hydration did not affect the overall capture capacity with an average of 880 μmol/g-sorbent over the four hydrated experiments compared to 890 μmol/g-sorbent with non-hydrated adsorption. A non-direct role of water in the adsorption of CO2 can be inferred when D2O and H2O were introduced and an isotopic shift was not observed in the carbonate IR bands.

Benzene probing: A C-C stretch at 1479 cm^{-1}, C-H out-of-plane vibration at 1950, 1854 and 1803 cm^{-1} became less intense after 15 minutes of flushing after the benzene was removed. The bands 1950 and 1854 cm^{-1} were observable after 15 minutes and 1803 cm^{-1} had disappeared after 15 minutes. The C-C stretch and C-H in-plane vibration at 3046 and 3092 cm^{-1} respectively, are associated with gas phase benzene and were not detectible after 15 minutes.

The untreated beta zeolite may have a liquid phase type interaction at the C-C bond level as from peaks evolved at 1480 and 3096 cm-1. However, the C-H stretch and out-of-plane vibrations are similar to vapor phase although they exhibit a shift that is not observed in gas phase. An acid-base interaction with the zeolite may have contributed to the shift since the benzene is a weak base and the zeolite is an acid; the amine is also a base hence a shift is not observed with the amine impregnated zeolite. The benzene peaks decay at a faster rate on the amine impregnated zeolite for all wavenumbers then with the zeolite suggesting the benzene has a stronger interaction with the untreated benzene than the amine impregnated benzene. The observation of adsorbed benzene on the zeolite and rate of disappearance on amine impregnated zeolite suggests a majority of the active acid sites are consumed by amines on the amine impregnated sample.

Conclusions:

The present study provides an insight on the mechanism of amine impregnated beta zeolite with in-situ IR spectroscopy and a mass spectrometer. CO_2 adsorbs as carbonates on the amine and a change was observed with and without hydration although the capture capacity was similar. The use of D_2O suggested there is role of water is non-direct via a lack of isotopic shifts. The observation of adsorbed benzene on the zeolite and rate of disappearance on amine impregnated zeolite suggests a majority of the active acid sites are consumed by amines on the amine impregnated sample.

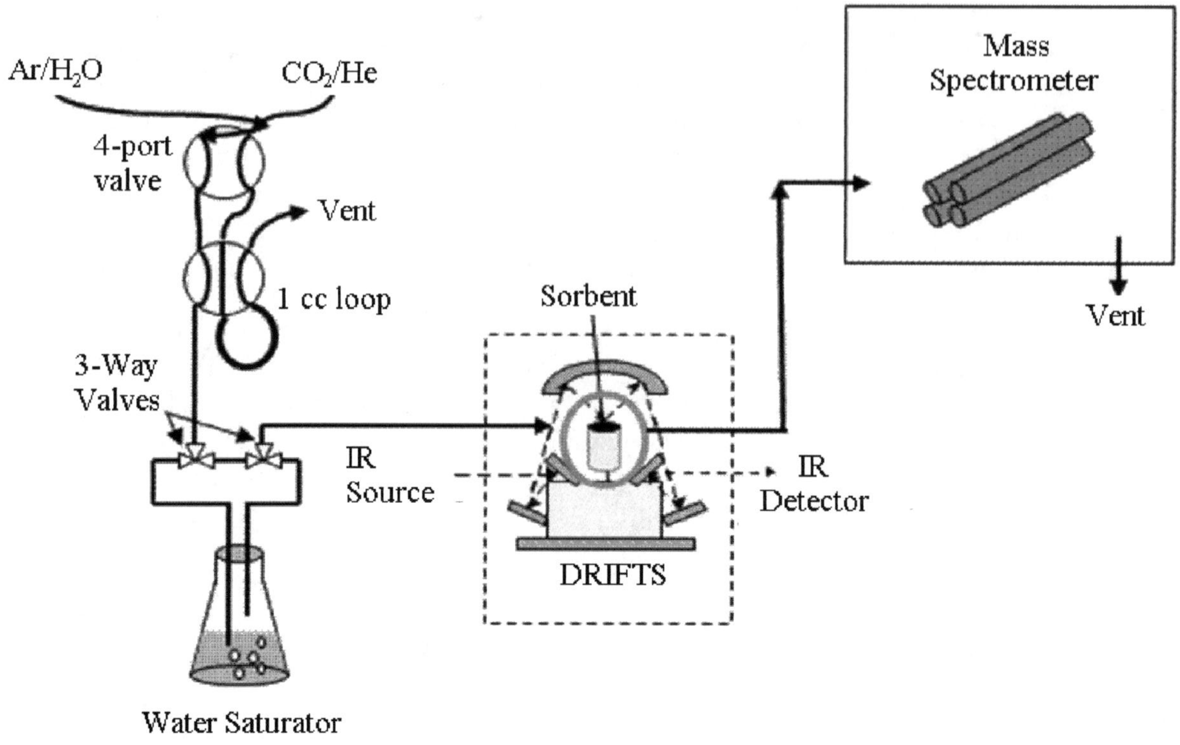

Fig. 1: The Experimental apparatus consists of a gas manifold which includes a 4-port valve and 6-port valve, a water saturator, a DRIFTS reactor filled with 40-60 mg of sorbent, and a mass spectrometer for effluent analysis.

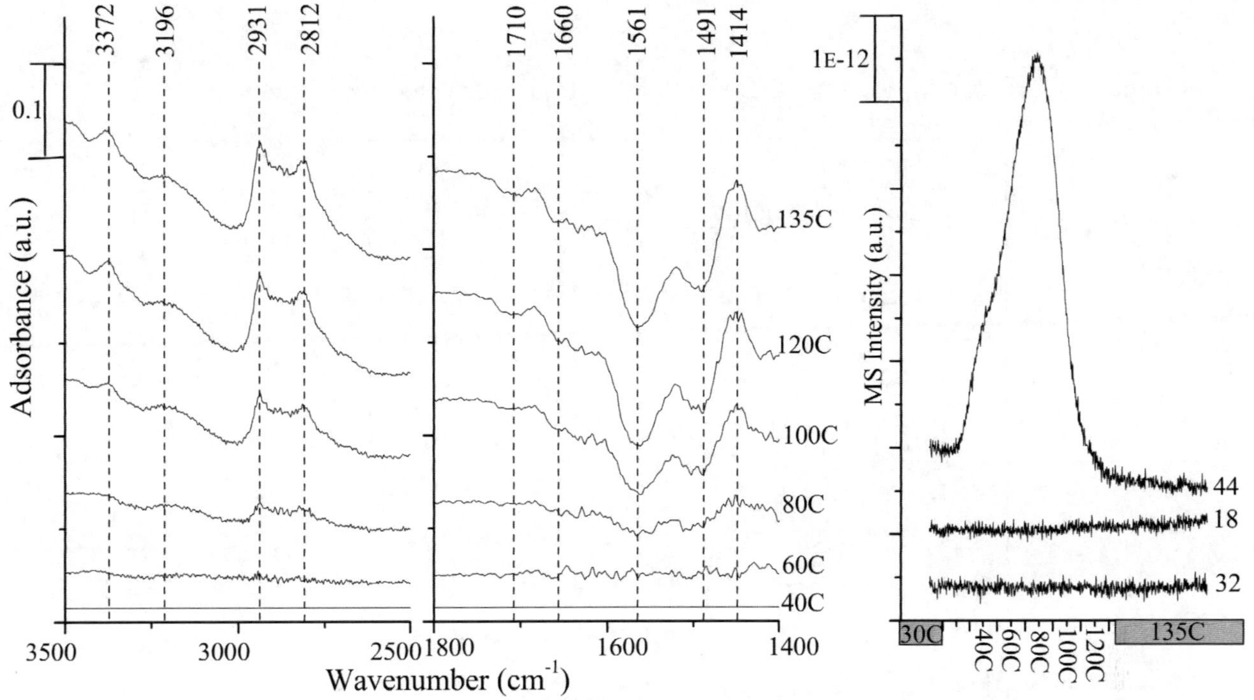

Fig. 2: Difference Spectra and MS profile during TPD after exposure to dry CO2 with a heating at a rate of 10C/min and flowing pure Ar at 30 cc/min.

Table 1: FTIR band assignments for amine, carbonate species and benzene.

Band Assignment	wavenumber (cm^{-1})
Amine	
combination of C-C and C-N	1050
C-N vibration	1144
C-H_2 wag	1266
C-H_2 deform	1460
N-H_2 sissoring	1597
N-H_2 sissoring	1650
C-H_2 vibration	2810
C-H_2 vibration	2930
N-H and N-H_2 vibration	3300
N-H and N-H_2 vibration	3372
Carbonate	
monodentate carbonate	1321
monodentate bicarbonate	1407
monodentate bicarbonate	1488
bidentate carbonate	1564
Benzene	
C-C stretch	1479
C-H out of plane	1803
C-H out of plane	1854
C-H out of plane	1950
C-H in plane and C-C strech	3046
C-H in plane and C-C strech	3092

68e A Generic Analysis of Energy Use and Solvent Selection for CO2 Separation from Post-Combustion Flue Gases

Yongqi Lu[1], Shiaoguo Chen[2], and Massoud Rostam-Abadi[1]. (1) Energy and Environmental Engineering, Illinois State Geological Survey, 615 E Peabody Drive, Champaign, IL 61820, (2) Science Applications International Corporation, 626 Cochrans Mill Road, Pittsburgh, PA 15236

Absorption-based processes are potentially the most attractive technologies for CO2 separation from post-combustion coal flue gas. The energy used in a typical absorption process has three major components: reaction heat (heat of absorption), stripping heat (water evaporation), and sensible heat (heating the solvent to the stripping temperature). The relative importance of these energy components depends on solvent type and operating conditions of the absorption and stripping columns. In a state-of-the-art mono-ethanol-amine (MEA) process, the reaction, stripping and sensible heat contribute 53%, 33% and 14% of the total heat use, respectively.

A thermodynamic calculation was performed to determine the theoretical minimum energy used to separate CO2 from a coal combustion flue gas in a typical adsorption-desorption system. The results showed that, under ideal conditions, the minimum energy required to separate CO2 from post-combustion flue gas and produce pure CO2 at 1 atmospheric pressure was only about 1,183 kJ/kg CO2. This amount could double with the inclusion of the driving forces of mass and heat transfer and the adverse impacts of absorption heat release on adsorption capacity.

Thermodynamic analyses also were performed for the aqueous amine-based absorption process. The heat requirement for hypothetical solvents covering a wide range of heat of absorption and the equilibrium constant of the CO2 reaction was examined. Two CO2 reaction mechanisms, the carbamate formation reaction with primary/secondary amines and the CO2 hydration reaction with tertiary amines, were included in the absorption reaction. The relation between the reaction equilibrium and the heat of reaction was examined for commercial amines. The results showed that the reaction heat, sensible heat and stripping heat are all important to the total heat requirement. The heat use of an ideal tertiary amine amounted to 2,786 kJ/kg, compared to 3,211 kJ/kg for an ideal primary amine. The heat usage of an ideal amine is about 20% lower than that of commercially available amines. Optimizing the absorption process configuration could further reduce energy use.

68f Evaluation of Ionic Liquids as Novel Post-Combustion Carbon Dioxide Capture Solvents

Edward J. Maginn[1], JaNeille K. Dixon[1], Elaine Mindrup[1], Wei Shi[2], Joan F. Brennecke[1], and William F. Schneider[1]. (1) Chemical and Biomolecular Engineering, University of Notre Dame, 180 Fitzpatrick, Notre Dame, IN 46556, (2) Chemical and Biomolecular Engineering, Univ. of Notre Dame, Notre Dame, IN 46556

Ionic liquids (ILs) are salts with low melting points, many below room temperature. Even though they are liquids, they have negligible vapor pressure. Typical ILs are composed of a imidazolium, pyridinium, ammonium or phosphonium cation with any of a wide variety of anions. Their properties can be varied tremendously by the choice of anion, cation and substituents. In research supported by the U.S. Department of Energy's National Energy Technology Laboratory and in collaboration with several industrial partners, we have shown that CO2 and SO2 both have remarkably high physical solubilities in some ILs. Since ILs are non-volatile, they have an advantage over conventional solvents used in carbon capture because they will not evaporate during absorption or regeneration. By making the substituent groups capable of forming a chemical complex with CO2, it is possible to greatly enhance capacity. By tuning the binding strength, the energy to regenerate the solvent can be minimized while maintaining high capacity.

In this talk, we will describe our recent progress in the development of new ionic liquids that can be used in carbon capture applications. Through a combined experimental and computational modeling effort, we have been able to greatly increase the solubility of CO2 in ionic liquid solvents. We report results of pure and mixed gas solubility and discuss preliminary process design and engineering studies that investigate the feasibility of using ionic liquids for post-combustion carbon capture.

68g Performance Evaluation of Aqueous Ammonia Process for CO2 Capture

Jong-Nam Kim[1], Kwang Bok Yi[1], You Jeong Kim[2], Chang Hyun Ko[1], Jong-Ho Park[1], and Won Hi Hong[2]. (1) Korea Institute of Energy Research, 71-2 Jang-dong Yuseong-gu, Daejeon, South Korea, (2) KAIST, 371-3, Guseong-dong Yuseong-gu, Daejeon, South Korea

Recently, aqueous ammonia process has been studied to reduce the cost of CO2 capture from flue gas of coal-fired power plant. In addition to its low price, ammonia process has several advantages, such as large CO2 sorption capacity, fast sorption rate, low regeneration energy and temperature, no degradation of absorbent by acidic compounds in flue gas, and potential ability to capture acidic gases. However, aqueous ammonia process has inherent problems due to the high vapor pressure of ammonia. In this study, the performance of continuous CO2 separation process using aqueous ammonia was evaluated. The best ammonia concentration was selected. CO2 transfer capacity, which can be defined as grams of CO2 recovered per gram of aqueous ammonia, and CO2 recovery were obtained in reasonable range of gas and liquid flow rates. In addition, the concentration changes of carbon containing species between absorption and regeneration were analyzed.

91a Evaluation of Potential Impacts of Geologic Carbon Sequestration Using Numerical Simulation

Brian McPherson[1], **Weon Shik Han**[2], Milind D. Deo[3], and Si-Yong Lee[2]. (1) Civil Engineering, University of Utah, 122 South Central Campus Drive, 104 CME, Salt Lake City, UT 84112, (2) Department of Civil and Environmental Engineering, University of Utah, 122 South Central Campus Drive, Salt Lake City, UT 84112, (3) Chemical Engineering, University of Utah, 50 S. Central Campus Dr., MEB, Rm. 3290, Salt Lake City, UT 84112

Long-term storage or sequestration of anthropogenic "greenhouse gases" such as CO2 is a proposed approach to managing climate change. Deep subsurface brine reservoirs in geological sedimentary basins are possible sites for sequestration, given their ubiquitous nature. We used a mathematical sedimentary basin model, including full coupling of multiphase CO2-groundwater flow, heat flow, and rock deformation, to evaluate effects of CO2 injection on fluid pressures and rock strain. We also analyzed residence times and migration patterns of CO2 in possible brine reservoir storage sites. Study areas include the Uinta and Paradox basins of Utah, and the Permian basin of west Texas. Regional-scale hydrologic and mechanical properties, including the presence of fracture zones, were calibrated using laboratory measurements and field data. Our initial results suggest that, in general, long-term (~100 years or more) sequestration in deep brine reservoirs is possible, if guided by robust rock mechanical and hydrologic data. However, specific processes must be addressed to characterize and minimize risks. In addition to CO2 migration from target sequestration reservoirs into other reservoirs or to the land surface, another environmental issue is displacement of brines into freshwater aquifers. We evaluated the potential for such unintended aquifer contamination by displacement of brines out of adjacent sealing layers such as marine shales. Results suggest that sustained injection of CO2 may incur significant brine displacement out of adjacent sealing layers, depending on the injection history, initial brine composition, and hydrologic properties of both reservoirs and seals. Model simulations also suggest that as injection can induce abnormally high fluid pressures, or overpressures. Results indicate that injection-induced overpressures may migrate, and effective stresses may follow this migration under some conditions, as will associated rock strain. Such "strain migration" may lead to induced or reactivated fractures or faults, but can be controlled through reservoir engineering.

Blow off limits with Oxy-Coal Combustion

Ari Roisman, Washington University in St. Louis, St. Louis, MO, USA
Melissa Holtmeyer, Washington University in St. Louis, St. Louis, MO, USA
Scott A. Skeen, Washington University in St. Louis, St. Louis, MO, USA
Richard L. Axelbaum, Washington University in St. Louis, St. Louis, MO, USA

Introduction

Oxy-coal combustion with flue gas recycle (FGR) is capable of achieving a flue gas with a dry basis CO_2 concentration of over 95% (Chen *et al.*, 2007), compared to 10-15% for traditional air-fired PC combustion. By producing a flue gas composed almost entirely of CO_2, oxy-coal combustion has the potential of enabling carbon capture and sequestration (CCS). Since proposed as a means to produce CO_2 for enhanced oil recovery, many laboratory-scale and pilot-scale studies of oxy-coal combustion have been performed. Studies have investigated the effects of oxy-coal combustion on NO_X reduction (Okazaki *et al.*, 1996, 1997; Hu *et al.*, 2000, 2001; Liu *et al.*, 2003, 2005; Croiset *et al.*, 2001), SO_X formation (Croiset *et al.*, 2001; Hu *et al.*, 2000; Liu *et al.*, 2001), char combustion (Shaddix and Murphy, 2003), heat transfer characteristics (Wall *et al.*, 2003), ignition characteristics (Kiga *et al.*, 1997), and the influence of swirl angles on flame stability (Nettleton, 2004). Oxy-coal with FGR has been shown capable of reducing NO_X emissions by as much as 60-70% (Châtel-Pélage *et al.*, 2006) and SO_X emissions by as much as 27% (Croiset and Thambimuthu, 2001) from the air-baseline case.

Of particular interest to the present study, Liu *et al.* (2005) found coal combustion in 21% O_2/79% CO_2 mixtures to exhibit delayed volatile release, pushing coal/char burning and associated bright luminosity further downstream than combustion in air. Because CO_2 has a larger specific heat than N_2, coal combustion in an O_2/CO_2 environment with component mass fractions comparable to those in air will exhibit lower gas temperatures and slower char burning, which tends to cause flames to destabilize and extinguish more easily than air-coal flames. To maintain a similar temperature distribution to that of coal combustion in air, the oxygen concentration in O_2/CO_2 mixtures needs to be increased to 30% (Liu *et al.*, 2005).

Oxy-Coal Combustion with Inert Exchange

Oxy-coal combustion studies have typically treated CO_2 as a replacement for N_2, which is used as a diluent to control flame temperature. Several gaseous fuel studies have shown that the placement of the inert in oxy-fuel combustion can have a significant impact on the combustion process, even if the flame temperature is held constant. For example, opposed-flow gaseous diffusion flame studies have shown that flame strength can be dramatically increased when the inert is removed from the oxidizer stream and introduced instead with the fuel stream (Chen and Axelbaum, 2005; Du and Axelbaum, 1996). Soot production can also be drastically reduced under these conditions (Du and Axelbaum, 1995, Kumfer *et al.,* 2005; Sunderland *et al.,* 2003).

Typically coal combustion is performed with air in both the coal and oxidizer streams. In conventional oxy-coal combustion both the coal stream and oxidizer streams contain identical ratios of O_2 to CO_2. In 'inert exchange' oxy-coal combustion, the O_2/CO_2 ratio can vary between the coal stream (primary oxidizer, PO) and secondary oxidizer (SO) streams. In this investigation, various O_2/N_2 and O_2/CO_2 coal flames were studied to better understand the effect of inert placement and inert type on flame stability, as indicated by the blow-off limit. Flow configurations for conventional air-coal combustion, conventional oxy-coal combustion, and inert exchange oxy-coal combustion are shown in Figure 1.

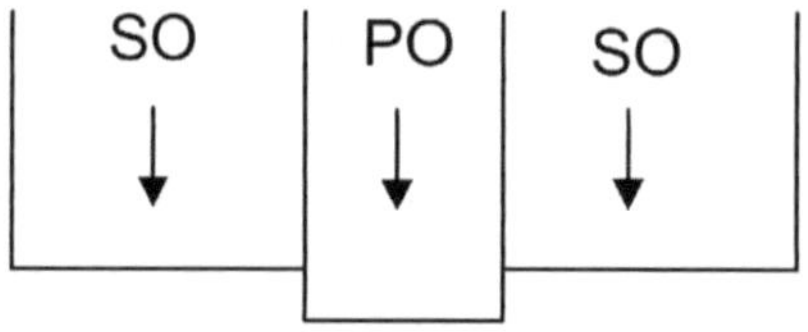

Coal-Gas Configuration	Primary Oxidizer (PO)	Secondary Oxidizer (SO)
Conventional Coal Air	Coal + Air	Air
Conventional Oxy-Coal	CO_2 + O_2	Coal + CO_2 + O_2
Inert Exchange Oxy-Coal	Diluent + (O_2)* + Coal	Diluent + O_2

* Oxygen mole fraction may be greater than or less than 21%

Figure 1. Flow configurations for conventional air- and oxy-coal, and inert exchange oxy-coal flame types

Experimental and Numerical

All studies were performed with a laboratory-scale oxy-coal combustor consisting of a downward-fired vertical combustor comprised of a non-swirling PO and tangentially-induced variable swirl SO capable of producing self-sustaining coal flames between 0.5 and 35 kW_{th}. Details can be found in Roisman (2007). Pulverized coal from the Black Thunder mine in Wyoming's Powder River Basin was used for this study.

Flame stability for coal flames of varying oxygen, nitrogen, and carbon dioxide concentrations was characterized by the mean velocity of the primary oxidizer at blow-off. The mean velocity is defined as the volumetric flowrate divided by the tube cross-sectional area. The system was operated by steadily increasing the volumetric flowrate of the primary oxidizer stream until blow-off was achieved. Blow-off was defined as the point at which the base of the flame was no longer visibly attached to the primary oxidizer tube and would not reattach. The mean PO velocity at the point of blow-off, V_{PO}, was calculated for each case at several coal feed rates corresponding to system power ratings between 17 and 35 kW_{th}. All of the SO flow was introduced tangentially and the SO flowrates were held constant for all cases at 6 scfm. The equivalence ratio based on the volatile component of the coal mixture in the PO was calculated from

$$\phi_{vol} = \frac{(V/O)}{(V/O)_{st}} \qquad (1)$$

where *V* represents the mass fraction of volatiles assuming all volatiles in the coal are released, and *O* represents the oxygen mass fraction.

To support the experimental effort, numerical simulations were performed using FLUENT 6.3. Effects of SO oxy-enrichment were simulated to determine PO component compositions required to match flame temperatures in the near burner region 25.4 mm downstream of the burner exit for inert exchange flame types compared with those for a conventional air-coal flame. Matching near-burner flame temperatures for the conventional air and inert exchange flame types allowed for comparison of flame behavior independent of near-burner temperature. Flame simulations were also used to help develop an understanding of flame volatility, structure, and temperature profiles for various flame types.

Results and Discussion

Numerical Simulation

Numerical simulations are discussed first to illustrate the effects of inert concentration in the various streams on flame temperature. Flames of varying levels of oxy-enrichment in both the PO and SO were studied. Three sample flame types from the numerical simulation, designated as flames A, B and C, are described in Table 1.

Table 1. Near burner region temperatures for air-, secondary oxy-coal, and inert exchange oxy-coal flames

Flame			*PO %**		*SO %**		T_{NB} *(K)*
Designation	*Type*	*Description*	O_2	N_2	O_2	N_2	
A	Air	Air in PO and SO	21	79	21	79	1860
B	Secondary Oxy	Oxy-enriched SO only	21	79	35	65	1950
C	Inert Exchange Oxy	Diluted PO, enriched SO	6	94	35	65	1860
*Primary Oxidizer (PO) and Secondary Oxidizer (SO) concentrations by molar content/volume							

Nitrogen was used as the inert in the simulations. Because the near burner temperature (T_{NB}) is an important parameter that can heavily influence flame stability, temperatures for each flame type were calculated and included in Table 1. For the secondary oxy-coal flame type, flame B, T_{NB} increased as oxygen was added to the SO. For the inert exchange flame type, flame C, T_{NB} was brought back down to the temperature of the air-coal flame by reducing the concentration of O_2 in the PO.

Near burner temperatures for flame B increased relative to flame A when the O_2 mole fraction in the SO was increased to 35%. By reducing the PO O_2 mole fraction to 6% while holding the SO O_2 mole fraction at 35%, T_{NB} for flame C was matched to T_{NB} for flame A. Other inert exchange conditions were also considered. For example, by reducing the PO O_2 to 15% while holding the SO O_2 at 25%, T_{NB} could be matched to T_{NB} for the air-coal flame. In this way, modeling was employed to identify inert exchange flame conditions that allow for comparative investigation with air-coal flames independent of near burner temperature.

Experimental

For each flame type, blow-off limits were characterized by a distinct change in luminous intensity at the burner exit. At the blow-off condition the flame was lifted such that burning occurred downstream of the burner exit. For most flames, blow-off was observed after a period of flame detachment and reattachment. As PO velocity was increased and the flames approached blowoff, the frequency of detachment and reattachment decreased until the flame would no longer reattach. For flames with the lowest oxygen concentrations and where the nitrogen was almost completely replaced with CO_2, the flames did not blow off but instead flame blow-out was observed, as characterized by complete flame extinction.

Table 2 shows the mole fractions of each component in the PO and SO for all flames studied. For cases in which CO_2 was run in the PO, some N_2 was still present in the PO due to the use of air instead of CO_2 in the eductor. This was done to prevent potential freezing associated with expansion.

Table 2. Flame types for experimental oxy-coal flame stability characterization study

Flame Type	*PO Mole Fraction*			*SO Mole Fraction*			T_{max}* *(K)*	T_B** *(K)*
%O_2 in PO/SO+	**O_2**	**N_2**	**CO_2**	**O_2**	**N_2**	**CO_2**	*from Fluent simulation*	
6%/35%	6	94	0	35	65	0	2420	1860
15%/25%	15	85	0	25	75	0	2440	1860
21%/21%	21	79	0	21	79	0	2480	1860
21%/25%	21	79	0	25	75	0	2600	1930
21%/35%	21	79	0	35	65	0	2850	1950
25%/25%	25	75	0	25	75	0	2720	1940
35%/21%	35	65	0	21	79	0		
35%/35%	35	65	0	35	65	0	3010	2070
21%/35% (N_2/CO_2)	21	79	0	35	0	65		
25%/25% CO_2++	25	23	52	25	0	75		
30%/30% CO_2++	30	21	49	30	0	70		
35%/21% CO_2++	35	23	42	21	0	79		
35%/35% CO_2++	35	25	40	35	0	65		

values for component compositions are accurate +/-2%
+ inert balance in both PO and SO is N_2 unless otherwise indicated
++ due to eductor, inert balance in PO is not entirely CO_2
* maximum temperature in temperature profile as determined from modeling
** flame temperature 1" downstream from burner exit as determined from modeling

A close correlation between blow-off limits for air-coal and 30% CO_2 oxy-coal flames was found and is shown in Figure 2. Results are supported by Liu *et al.* (2005) who investigated gas temperature profiles for various N_2 and CO_2 oxy-coal flames in a 20 kW down-fired combustor similar to the down-fired combustor implemented in this study. In that study, similar gas temperature profiles were measured for 30% O_2/CO_2 flames and air-coal flames at incremental axial distances from the burner. Croiset and Thambimuthu (2001) found that 35% O_2 with a balance of CO_2 yielded similar temperature profiles to conventional air combustion. Discrepancies between the two studies in terms of the oxygen concentration required to match heat transfer characteristics were attributed to the 10% entrained air present in all tests performed by Liu *et al.* (2005). Discrepancy between results presented here and those of Croiset and Thambimuthu (2001) could be attributed to the ~20% nitrogen present in the PO due to the method of coal entrainment mentioned above.

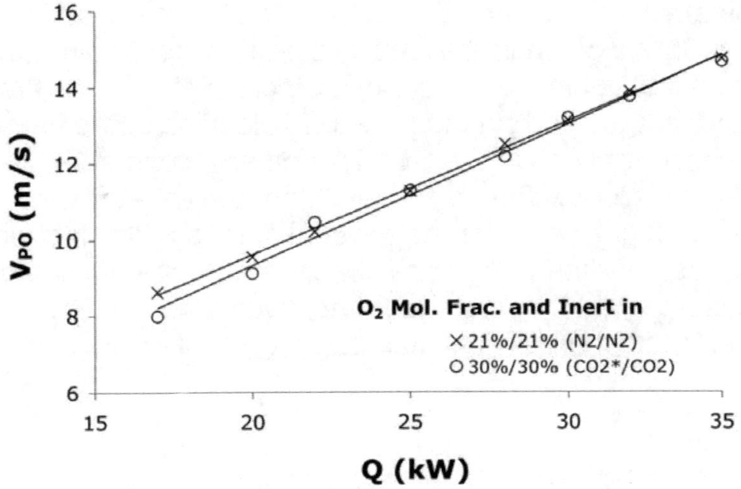

Figure 2. PO blow-off velocity for air and 30% O_2/CO_2 oxy-coal flames vs. thermal power

Figure 3 compares velocities at blow-off for nitrogen and carbon dioxide flame types. Blow-off results for flame types with 25%/25%, 35%/21%, and 35%/35% show a clear shift to lower velocities when carbon dioxide replaces nitrogen. The effect is most dramatic for the 25%/25% flame types, as these have the largest amount of inert. This effect of replacing N_2 with CO_2 is indicated by an arrow connecting N_2 flames to CO_2 flame for each flame type.

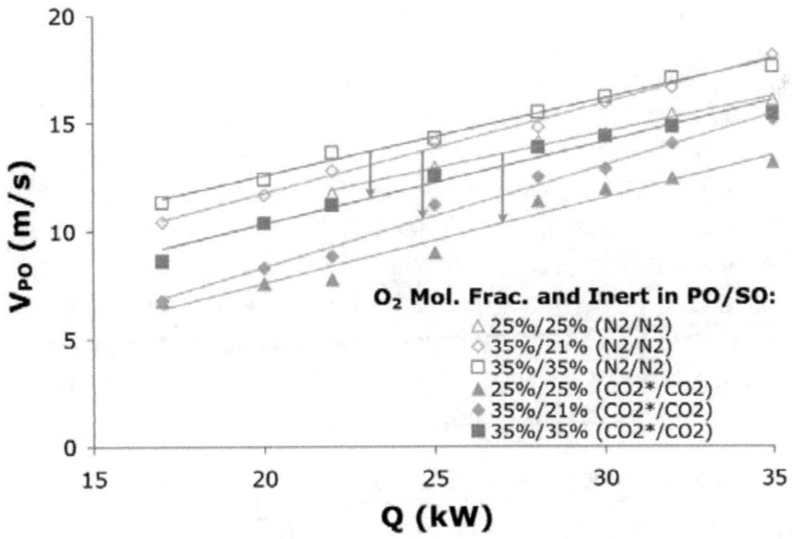

Figure 3. PO blow-off velocity for CO_2 and N_2 oxy-coal flames vs. thermal power

Inert exchange flames were produced by increasing the oxygen concentration of the SO while adding nitrogen to the PO. Inert exchange flames are compared with air-coal flames having the same local temperature to determine how structural changes in the flame might influence localized phenomena such as blow-off limits.

Figure 4 shows blow-off limits for two inert exchange flame types where the PO/SO oxygen concentrations are 6%/35% and 15%/25% as well as blow-off limits for flames with air in the PO and 25% and 35% oxygen in the SO. Blow-off limits for air-coal flames are also included for reference. As expected, the blow-off velocities for the inert exchange flames (where the PO is diluted) are lower than the blow-off velocities of the flames with air in the PO and elevated oxygen concentrations in the SO. This difference is attributed to lower near burner temperatures in the inert-exchange flames. Nonetheless, the blow-off velocities for the inert-exchange flames are higher than for the air-coal flames at the higher powers, even though local temperatures near the burner are similar to the air-coal flames. Figure 4 indicates, for most cases, that inert exchange flames with a diluted PO and oxy-enriched SO exhibited higher PO velocities than air-coal flames at the same power, indicating improved flame stability.

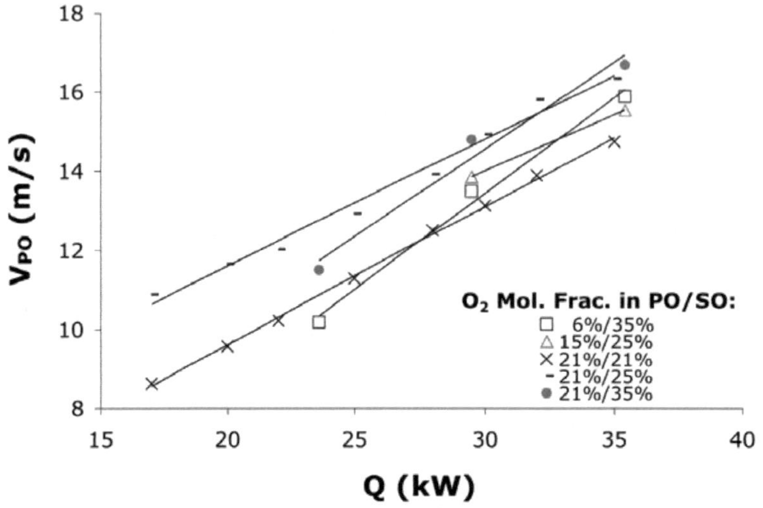

Figure 4. PO blow-off velocity for N_2 inert exchange coal flames vs. thermal power

The equivalence ratio assuming complete volatile release, ϕ_{vol}, was identified as a variable that could influence flame stability. The actual value of the equivalence ratio for the volatiles near the base of the flame are expected to be lower because complete devolatilization does not occur until some point downstream of the burner exit, but ϕ_{vol} is relevant for comparative purposes. It follows that flames with the highest temperatures are likely to more closely coincide with ϕ_{vol} data and flames with the lowest temperatures are most likely to exhibit greatest deviation from ϕ_{vol} values as presented.

Figure 5 shows the effect of substituting carbon dioxide for nitrogen while maintaining oxygen concentrations in both the PO and SO for 35%/35% and 35%/21% flame types. Blow-off velocity at a given equivalence ratio is higher in the nitrogen flames than in the carbon dioxide flames and comparable blow-off velocities are achieved at lower equivalence ratios for the nitrogen flames. As seen in Figure 5, blow-off velocities for the air-coal flame at a given power nearly match those for the 30%/30% CO_2 flame, despite a dramatic shift in equivalence ratio.

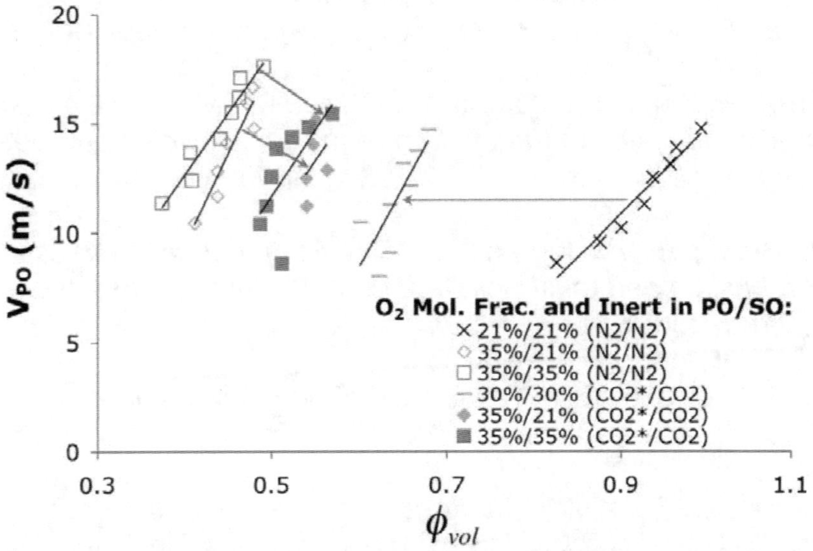

Figure 5. PO blow-off velocity for N_2 and CO_2 oxy-coal flames vs. ϕ_{vol}

Acknowledgements

The authors acknowledge the financial support of the United States Department of Energy under the University Coal Research program as well as AmerenUE and the Missouri Department of Natural Resources.

References

1. Châtel-Pélage F., Varagani R. *et al.*, *Applications of oxygen for NO_X control and CO_2 capture in coal-fired power plants.* Thermal Science; 10: 119-42 (2006).

2. Chen J.C. and Huang J.S., *Combustion efficiency and CO_2 emission from O_2/N_2, O_2/CO_2, and O_2/RFG coal combustion processes.* Environmental Engineering Science; 24(3): 353-62 (2007).

3. Chen R. and Axelbaum R.L., *Scalar dissipation rate at extinction and the effects of oxygen-enriched combustion.* Combustion & Flame; 142(1-2): 62-71 (2005).

4. Croiset E. and Thambimuthu K.V., *NO_X and SO_2 emissions from O_2/CO_2 recycle coal combustion.* Fuel; 80: 2117-21 (2001).Du J. and Axelbaum R.L., *The effect of flame structure on soot-particle inception in diffusion flames.* Combustion and Flame; 100(3): 367-75 (1995).

5. Du J. and Axelbaum R.L., *The effects of flame structure on extinction of CH_4-O_2-N_2 diffusion flames.* Proceedings of the Combustion Institute; 26(1): 1137-42 (1996).

6. Hu Y., Naito S., Kobayashi N., Hasatani M., CO_2, NO_X and SO_2 emissions from the combustion of coal with high oxygen concentration gases. Fuel; 79(15): 1925-32 (2000).

7. Hu Y.Q., Kobayashi N., Hasatani M., *The reduction of recycled-NO_X in coal combustion with O_2/recycled flue gas under low recycling ratio.* Fuel; 80(13): 1851-5 (2001).

8. Kiga T., Takano S., Kimura N., Omata K., Okawa M., Mori T., *Characteristics of pulverised-coal combustion in the system of oxygen/recycled flue gas comustion.* Energy Conversion Management; 38: 129-34 (1997).

9. Kumfer B.M., Biswas P., Axelbaum R.L., *Oxy-Combustion: Novel strategies for improving combustion and multi-pollutant control.* Annual International Pittsburgh Coal Conference, 22nd (2005).

10. Liu H., Katagiri S., Okazaki K., *Drastic SO_X removal and Influences of Various Factors in O_2/CO_2 Pulverized Coal Combustion System.* Energy & Fuels; 15: 403-12 (2001).

11. Liu H. and Okazaki K., *Simultaneous easy CO_2 recovery and drastic reduction of SO_X and NO_X in O_2/CO_2 coal combustion with heat recirculation*. Fuel; 82: 1427-36 (2003).

12. Liu H., Zailani R., Gibbs B.M., *Comparisons of pulverized coal combustion in air and in mixtures of O_2/CO_2.* Fuel; 84: 833-40 (2005).

13. Nettleton M.A., *The influence of swirl angles on flame stability in pilot-scale plant.* Fuel; 83: 253-6 (2004).

14. Okazaki H., Taniguchi M., Kobayahi H., Amano K., Azuhata S., Hishinuma Y., *Kinetics of devolatilization of pulverized coal ignited by pulsed laser*. Nippon Kikai Gakkai Ronbunshu, B-hen; 62(598): 2466-71 (1996).

15. Okazaki K. and Ando T., *NO_X reduction mechanism in coal combustion with recycled CO_2.* Energy; 22(2): 207-15 (1997).

16. Roisman, A., *Flame Stability in Oxy-Coal Combustion,* Master's thesis, Washington University, Department of Energy, Environmental & Chemical Engineering, St. Louis, MO (2007).

17. Shaddix C.R. and Murphy J.J., *Coal char combustion reactivity in oxy-fuel applications.* Proceedings – Annual International Pittsburgh Coal Conference, 20th: 909-23 (2003).

18. Sunderland P.B., Axelbaum R.L., Urban D.L., Chao B.H., Liu S., *Effects of structure and hydrodynamics on the sooting behavior of spherical microgravity diffusion flames.* Combustion and Flame; 132(1/2): 25-33 (2003).

19. Wall T.F., Gupta R., Sheng C.D., *Calculations of the heat transfer changes for O_2/CO_2 oxyfuel combustion rerofits of air fired Pf plant with associated recycle ratios and O_2 concentrations fed to the boiler.* University of Newcastle; Chemical Engineering, NSW 2308 (2003).

91c Oxygen Sorbents for Oxy-Fuel Combustion and Igcc Power Plants

Gokhan Alptekin, Ambalavanan Jayaraman, Margarita Dubovik, and Lauren Brickner. TDA Research Inc., 12345 W. 52nd Ave, Wheat Ridge, CO 80033

Oxy-fuel combustion has the potential to dramatically decrease the environmental impact of producing electrical power, especially because it can be retrofit to existing pulverized coal power plants. In this technology, oxygen combustion replaces the traditional combustion environment of coal in air. This results in an off gas of steam and pure CO_2, which can be easily sequestered rather than released into the atmosphere where it contributes to the green house effect. However, the technology is not yet used in spite of its promise, primarily because pure oxygen is currently too expensive. TDA Research, Inc. (TDA) is developing a low cost, high capacity sorbent for air separation to supply low cost oxygen for oxy-fuel combustion. The regenerable sorbent can remove oxygen from air with very high capacity, and the sorbent maintains its capacity for over 200 absorption/regeneration cycles under representative conditions. TDA's oxygen sorbents also would find application in IGCC plants. Using oxygen instead of air in the IGCC process will remove excess nitrogen from the gasifier and there by increase efficiency, reduce NOx emissions and improve gas purity. Removing nitrogen from the fuel gas is particularly critical if the synthesis gas is used for the production of chemicals instead of electricity.

91d Chemical Looping Combustion of Coal Derived Synthesis Gas Over Copper-Based Oxygen Carrier

Hanjing Tian[1], Ranjani V. Siriwardane[2], Goetz Veser[3], Thomas Simonyi[4], Tengfei Liu[3], and Rahul Solunke[5]. (1) Parsons Infrastructure & Technology Group, Inc, 3610 Collins Ferry Road, P.O. Box 880,, Morgantown, WV 26505, (2) National Energy Technology Lab/ U.S. Department of Energy, 3610 Collins Ferry Road, Morgantown, WV 26507-0880, (3) Department of Chemical Engineering, University of Pittsburgh, 1249 Benedum Engineering Hall, Pittsburgh, PA 15261, (4) Research Development Solutions, 3610 Collins Ferry Road, P.O.Box 880, Morgantown, WV 26507, (5) Chemical Engineering Department, University of Pittsburgh, 1249 Benedum Hall, Pittsburgh, PA 15261

Chemical looping combustion (CLC) is an emerging technology for clean energy production from fossil and renewable fuels. In CLC, an oxygen carrier (typically a metal) is first oxidized with air. The hot metal oxide is then reduced in contact with a fuel in a second reactor, thus combusting the fuel. Finally, the reduced metal is transferred back to the oxidizer, closing the materials "loop". CLC thus allows for flame-less, NOx-free combustion without requiring expensive air separation. Most importantly, CLC produces sequestration-ready CO2-streams without significant energy penalty. However, CLC is currently suffering from insufficient stability of the oxygen carrier (particularly metal sintering), and slow metal re-oxidation kinetics.

We present results from a study in which CuO/bentonite and nanostructured Cu-Barium-Hexaaluminate (Cu-BHA) oxygen carriers were studied in chemical looping combustion for the gasification systems utilizing simulated synthesis gas. Global reaction rates of reduction and oxidation, as the function of reaction conversion, were calculated from 10-cycle oxidation/reduction tests utilizing thermogravimetric analysis at atmospheric pressure between 700 °C to 900 °C. It was found that the reduction reactions are always faster than oxidation reaction; reaction temperature and particle size do not significantly affect the reaction performance of CuO/bentonite. Results of the tapered element oscillating microbalance showed negative effect of pressure on the global rates of reduction-oxidation reactions at higher fractional conversions. X-ray photoelectron spectroscopy analysis of fresh and used CuO/bentonite revealed the presence of multiple oxidation states in the surface of oxidized sample and metallic Cu in the surface of reduced sample. X-ray diffraction patterns confirmed the presence of CuO in the bulk phase of oxidized sample. Scanning electron microscopy analysis showed significant morphology changes of reacted CuO/bentonite samples after the 10 oxidation-reduction cycles above 700oC in an atmospheric thermogravimetric analyzer.

The performance of this bentonite-based carrier was compared with that of a nanostructured Cu-BHA carrier. The Cu-BHA carrier was synthesized following a microemulsion-templated synthesis path, and characterized by BET, XRD, and TEM. It was then subjected to the same 10-cycle redox treatment as the bentonite carriers. We find that the nanosized Cu particles in the Cu-BHA are highly stable at the high-temperature conditions of the experiment due to a 'caging' effect by the mesoporous BHA matrix. A detailed analysis of the oxidation and reduction kinetics of the two carriers, based on a shrinking core model, showed that nanostructuring of the carrier does not lead to significant acceleration of the redox kinetics. This is in stark contrast to our earlier findings with Ni-based carriers, where nanostructuring was found to have a strong accelerating effect on the kinetics. The difference can be explained based on the (solid state) diffusion rates for the two oxides.

Overall, we find that both CuO/bentonite and Cu-BHA exhibit excellent reaction performance and thermal stability for CLC process at 700-900 oC.

ANALYSIS OF HOT/WARM CO_2 CAPTURE FOR IGCC PROCESSES

Jared Ciferno[1], Shiaoguo Chen[2] and Wen-Ching Yang[2], [1]U.S. Department of Energy, National Energy Technology Laboratory, P.O. Box 10940, Pittsburgh, PA 15236-0940, [2]Science Applications International Corporation, 626 Cochrans Mill Road, Pittsburgh, PA 15236

Introduction

There is a growing concern that the buildup of CO_2 and other greenhouse gases in the atmosphere is contributing to climate change with undetermined consequences. To address this problem, the Department of Energy's National Energy Technology Laboratory (NETL) has been developing technologies that will allow coal, our most abundant fossil fuel, to be utilized for power generation with very low emissions of CO_2. One promising approach is the Integrated Gasification Combined Cycle (IGCC), which gasifies the coal to produce syngas, subject the syngas to the water-gas-shift (WGS) reaction for conversion of CO to CO_2 and H_2, remove the CO_2, and send the H_2 fuel to a combustion turbine. The recovered CO_2 is compressed and sequestered. Figure 1 is the schematic diagram of an IGCC process. For IGCC to be attractive, an efficient and economical CO_2 capture process is necessary.

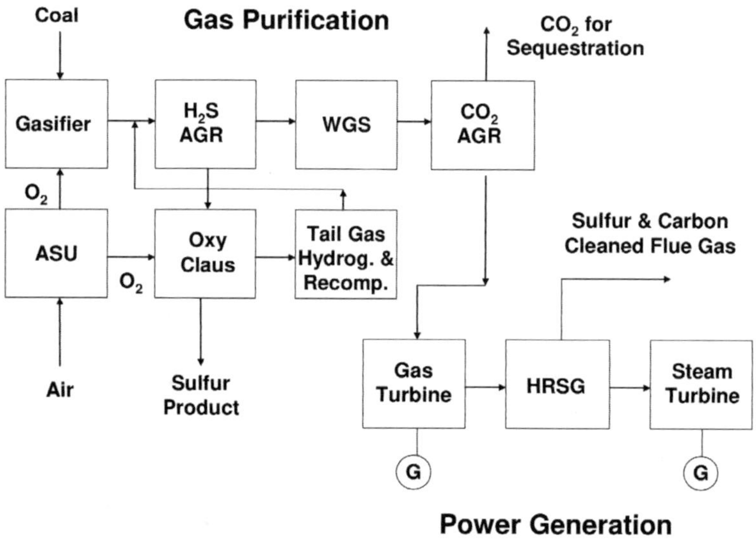

Figure 1. Schematic diagram of an IGCC process with CO_2 capture

Pre-combustion CO_2 capture is preferred for IGCC power plants due to the fact that gasification systems operate under high pressure, which makes CO_2 capture equipment much smaller in size compared to the post-combustion configuration.

To remove all the carbon in syngas as CO_2, WGS reaction is employed to convert CO to CO_2:

$$H_2O + CO == CO_2 + H_2 + \Delta H \quad (-41.2\ kJ) \tag{1}$$

The WGS reaction is usually carried out at temperatures between 250~450°C (450 to 800°F), depending on the WGS catalyst used. The WGS reaction is slightly exothermic so low temperature favors the forward reaction. After the WGS process, the CO_2 partial pressure in the syngas stream is significantly increased and will greatly facilitate CO_2 removal.

After the syngas is shifted, CO_2 is removed and then the hydrogen-rich syngas is sent to a gas turbine to generate electricity. The typical syngas conditions entering gas turbine are at around 260°C (500°F) and 30.6 atm (450 psi). It is desirable to remove the CO_2 at a temperature between the WGS reaction temperature and gas turbine inlet temperature. Such a CO_2 removal process will eliminate cooling/heating of the syngas and thus improve the thermal and economic performance of the IGCC process. This is exactly the incentive for developing CO_2 removal processes that operate at high temperatures.

A generic separation process for CO_2 capture from shifted syngases

Many separation processes are available for CO_2 removal from shifted syngas. Absorption, adsorption, cryogenic (low temperature distillation) and membrane processes are all possible options. However, since the focus here is on the hot/warm CO_2 removal, conventional absorption processes liquid type solvents (such as Selexol^{TM}) and cryogenic process will not be considered. Membrane processes, on the other hand, only use mechanical power and will not impact the steam cycle or gas turbine cycle of the power plant and will not be analyzed as well.

Assumptions for the generic adsorption process

Consider a conceptual adsorption process (see Figure 2 for schematic flow diagram) with the following reversible reaction:

$$M + CO_2 \Leftrightarrow N - \Delta H \qquad (2)$$

Here M is the sorbent, N is the reaction product and ΔH is the heat of adsorption.

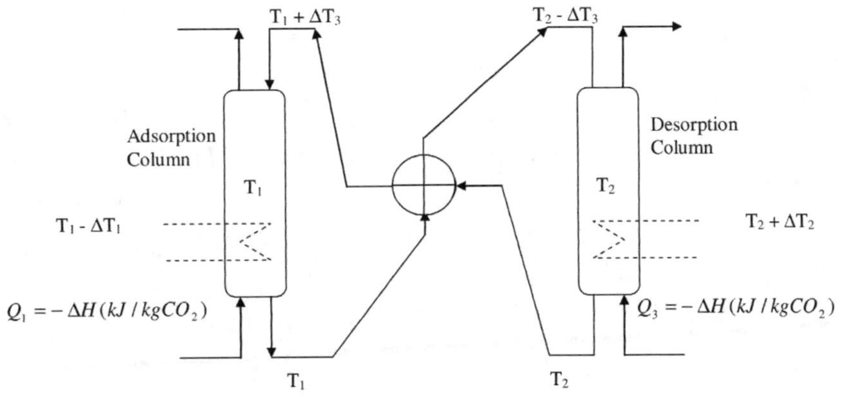

Figure 2. Schematic diagram of a generic adsorption process

Since our interest is in hot/warm CO_2 capture processes, it is reasonable to assume that the adsorption process is chemical in nature, that is, the ΔH of reaction 2 is high compared with the ΔH of physical adsorption. More importantly, because ΔH is high a temperature swing adsorption cycle will be the most appropriate and will be adopted here for this analysis.

Assuming the adsorption/desorption cycle has the following operating parameters:

Adsorption (working) capacity	= W ($kgCO_2$/kg sorbent)
Heat of adsorption	= ΔH ($kJ/kgCO_2$)

CO_2 adsorption temperature	$= T_1$ (K)
Regeneration (desorption) temperature	$= T_2$ (K)
CO_2 partial pressure at adsorption	$= P_1$ (atm.)
CO_2 partial pressure at desorption	$= P_2$ (atm.)
Heat capacity (for both M and N)	$= C_p$ (kJ/K/kg sorbent)
CO_2 recovery rate	= 90%

Using these assumptions and some simplifications, the heat effects of a conceptual adsorption/desorption cycle can be calculated.

Energy consumption of the generic adsorption process

Consider the adsorption/desorption cycle as depicted in Figure 3.

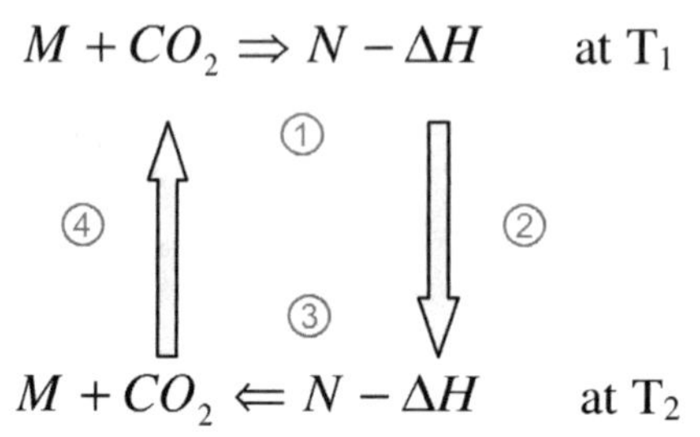

Figure 3. The conceptual adsorption/desorption cycle

In step ①, the heat generated is Q_1 and has a temperature of T_1:

$$Q_1 = \Delta H \, (kJ / kgCO_2)$$

In step ②, Q_2 is required to heat the spent sorbent to T_2 and can be calculated:

$$Q_2 = C_p / W \times (T_2 - T_1) \quad (kJ / kgCO_2)$$

The heat required in this step is sensible heat which is evenly distributed in the temperature range of T_1 to T_2.

In step ③, the heat required for regeneration of the spent sorbent is the same as in step ①; assuming that ΔH is constant (independent of temperature), which is:

$$Q_3 = -\Delta H \quad (kJ / kgCO_2)$$

Usually, ΔH and heat capacity C_p are a weak function of temperature. For simplicity, this analysis assumed they are independent of temperature and should hold well within a certain range of temperature.

In step ④, the regenerated sorbent will be cooled from T_2 to T_1 and the heat liberated can be calculated as follows (again, C_p is assumed to be constant):

$$Q_4 = C_p / W \times (T_1 - T_2) \quad (kJ / kgCO_2)$$

Q_2 and Q_4 are both sensible heat and can be integrated to save energy. However, a ΔT_3 is required between these two streams, that is, the regenerated sorbent (at T_2) can only heat the spent sorbent to T_2-ΔT_3. Therefore, the *net heat required* to heat the spent sorbent from T_1 to T_2 is:

$$Q_s = C_p / W \times \Delta T_3 \quad (kJ / kgCO_2)$$

All the heats liberated/consumed in the adsorption cycle can be integrated into the steam cycle. However, a ΔT is required to transfer heat from the sorbent cycle to steam cycle and vice versa. The steam generated from step ① will have a temperature of T_1-ΔT_1 and steam consumed in step ③ will have a temperature of $T_2 + \Delta T_2$ respectively. And steam necessary to provide the sensible heat Q_s will also have a temperature $T_2 + \Delta T_2$ as well.

To summarize, the heat consumed in the cycle step ② and ③ is:

$$Q_T = Q_3 + Q_s = \Delta H + C_p / W \times \Delta T_3 \quad (kJ / kgCO_2 \quad at\ T_2 + \Delta T_2)$$

And the heat recovered in step ① and ④ is:

$$Q_R = Q_1 + Q_s = \Delta H + C_p / W \times \Delta T_3 \quad (kJ / kgCO_2 \quad at\ T_1 - \Delta T_1)$$

Important to note that although Q_T and Q_R are the same quantity, the heat is different quality.

Now since the heat requirement for the whole cycle is known, the heat can be integrated into steam cycle to convert to electricity if we know the thermal efficiency of the heat.

Thermal Efficiency of Steam

It is difficult to assign a thermal efficiency to heat at any temperature. The conversion efficiency of heat to electricity is a function of heat quality (temperature) and conversion technology used. Different technologies will give different thermal efficiencies. For example, the same amount of heat may produce different amounts of electricity depending on the technologies used or even with the same technology it may generate different amounts of electricity due to different operating parameters. Owing to these facts, the conversion processes for this analysis was restricted to a typical a steam cycle and the conditions of the steam will be on the unique expansion path (adiabatic expansion path) and will end at the same conditions (typical cooling tower). With these two restrictions heat at any temperature will have a unique thermal efficiency to produce electricity. Here is how the thermal efficiency was estimated for this analysis.

First, heat is used to generate steam on the unique expansion path and then the steam is expanded in the steam cycle to produce electricity. The thermal efficiency of the steam is taken as the thermal efficiency of heat. Figure 4 shows the thermal efficiency of heat as a function of temperature generated by CHEMCAD.

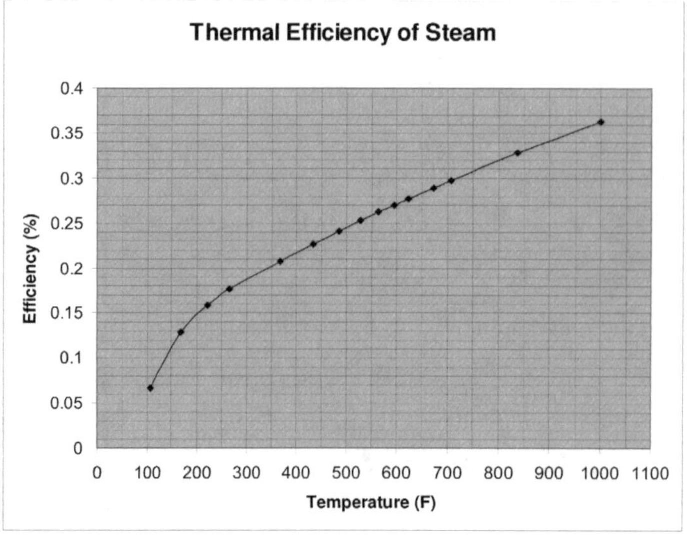

Figure 4. Steam properties during expansion in the IP and LP turbines

Electricity usage of hot/warm CO_2 removal processes

Knowing the thermal efficiency of steam the electricity usage of the hot/warm adsorption processes can be calculated.

Assuming η_1 and η_2 are thermal efficiency of steam at T_1-ΔT_1 and T_2+ΔT_2 respectively. The total electricity usage of the adsorption cycle is then:

$$E = \eta_2 \times Q_T - \eta_1 \times Q_R = (\Delta H + C_p / W \times \Delta T_3) \times (\eta_2 - \eta_1) \quad (kJ / kgCO_2) \tag{3}$$

Given heat of adsorption, heat capacity of the sorbent, working capacity of the sorbent and temperature difference required for heat transfer as well as the thermal efficiency of the steam, we will be able to calculate the electricity usage of an adsorption process by equation 3.

It should be pointed out that work to compress the CO_2 to pipeline specifications will be added if P_2 is lower than 2200 psi, which is the assumed pipeline pressure specification used here.

Thermodynamics of the hot/warm CO_2 removal

Hot/warm CO_2 removal processes have the potential to improve the thermal efficiency of an IGCC power plant and also reduce capital cost by eliminating heating and cooling of the syngas. In this section, a general hot/warm CO_2 removal process is analyzed from a thermodynamics perspective to see what kind of heat of adsorption is required for IGCC applications.

Consider reaction (2):

$$M + CO_2 \Rightarrow N + \Delta H \tag{2}$$

For simplicity, both M and N are assumed to be solids (e.g. oxide/carbonate cycle). The equilibrium constant of reaction (2) and its relationship to other thermodynamic variables are

$$K = 1/P_{CO2} \tag{4}$$

$$\Delta G = \Delta H - T\Delta S = -RT \ln K = -RT \ln \ (1/P_{CO2}) = RT \ln P_{CO2} \tag{5}$$

$$\Delta H = \Delta G + T\Delta S = RT \, Ln \, P_{CO2} + T\Delta S \tag{6}$$

$$\Delta S = S_N - S_M - S_{CO2} \tag{7}$$

The entropy of many oxides and their carbonates are listed in Table 1. Typically, the entropy difference between oxide and its carbonate is in the range of 38 to 64 (J/K/mol).

Table 1. Entropy of the common molecules seen in CO_2 capture [1]

Compounds	ΔH_f (kJ/mol)	ΔG_f (kJ/mol)	S (J/ K/mol)	S_N - S_M
CO_2(g)	-393.5		213.6	
CO_2(aqueous)	-412.91		121.3	
Li_2O	-142.4		37.9	52.5
$Li2CO_3$	-1215.6	-1132.4	90.4	
Na_2O	-415.9	-376.6	72.8	63.2
Na_2CO_3	-1130.9	-1047.7	136.0	
K_2O	-361.5		96.2	59.8
K_2CO_3	-1151.0	-1063.5	156.0	
MgO(S)	-601.2	-569.6	26.8	38.9
$MgCO_3$	-1116.0	-1029.3	65.7	
CaO(s)	-635.1	-604.2	39.8	53.1
$CaCO_3$(s)	-1206.9	-1128.8	92.9	
SrO	-590.4	-559.8	54.4	42.7
$SrCO_3$	-1218.4	-1137.6	97.1	
BaO	-558.1	-528.4	70.3	41.8
Ba_2CO_3	-1218.8	-1138.9	112.1	
CuO	-155.2	-127.2	43.5	44.4
$CuCO_3$	-595.0	-518.0	87.9	
PbO	-219.2	-189.3	68.2	62.8
$PbCO_3$	-700.0	-626.3	131.0	
Ag_2O	-30.6	-10.8	121.7	45.7
Ag_2CO_3	-506.1	-437.1	167.4	
ZnO	-348.0	-318.2	43.9	38.5
$ZnCO_3$	-812.5	-731.4	82.4	
Range				**38 - 64**

Substituting into equation (6):

$$\Delta H = RT\ln P_{CO2} + T\ (S_N - S_M - S_{CO2}) = RT\ln P_{CO2} - (149.6\text{\textasciitilde}175.6)\ T$$

Using typical CO_2 partial pressure in syngas (P_{CO2}=16atm, P_{total} = 40atm and 40% concentration) and assuming a 90% removal we have P_{CO2} =1.6 atm and:

$$\Delta H = RT\ \ln(1.6) - (149.6\text{\textasciitilde}175.6)\ T = -\ (145.7\text{\textasciitilde}171.7)\ T \tag{7}$$

This is the heat of adsorption required for a typical oxide/carbonate based adsorption process. Figure 5 is a graphical representation of the above equation.

At the desired warm temperature between WGS and gas turbine, e. g. 500°F (260°C):

$$\Delta H = -\ (145.7\ \text{\textasciitilde}171.7)*533.15 = -\ (77.7\text{\textasciitilde}91.5)\ (kJ/molCO_2)$$

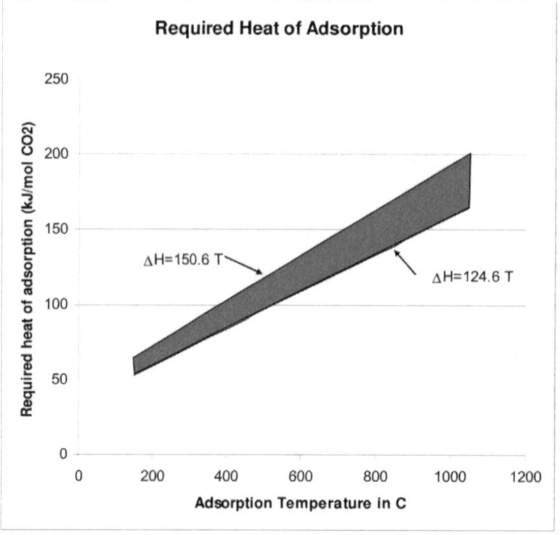

Figure 5. Required heat of adsorption for conceptual hot/warm CO_2 removal processes

For simplicity, we neglect the sensible heat portion ($C_p / W \times \Delta T_3$) in equation (3) and further assume T_2=T_1, $\Delta T_2 = 0$ (See Figure 2). In a real power plant, heat at 500F may not have much economic value to be injected into steam cycle and the heat released during adsorption step may not be recovered, that is, η_1=0 and η_2=0.25 (from Figure 4, at 500F) and equation (3) becomes:

$$E = (\Delta H + C_p / W \times \Delta T_3) \times (\eta_2 - \eta_1) \approx \Delta H \times (\eta_2 - \eta_1) = (77.7 \sim 91.5) \times 0.25\ (kJ / molCO_2)$$
$$= (19.4 \sim 22.8)\ (kJ / molCO_2) = (441 \sim 520)\ (kJ / kgCO_2)$$

This electricity usage is much higher than the electricity usage of the state-of-the-art Selexol™ scrubbing process (187kJ/kgCO_2). In addition, if the desorbed CO_2 pressure is low, then compression work will have to be considered which makes the warm CO_2 removal process even harder to compete with Selexol™ process.

Suggested processes for CO_2 removal from shifted syngas

Above discussion showed that hot/warm adsorption based CO_2 removal processes require a high heat of adsorption and is energy intensive. A lower temperature solvent-based CO_2 removal process may be more attractive not only because it requires much lower heat of absorption but also because liquid is much easier to deal with than solids.

When a liquid solvent (physical or chemical) is used, the minimum heat of absorption will not exist since the entropy of mixing is involved and ΔS of reaction (2) is a function of the conversion rate of M to N. However, we may get some idea of a typical heat of absorption required from the entropy of CO_2 in aqueous liquid.

The entropy of CO_2 in water is 121.3 (J/K/mol). Assuming M is water, and N is a mixture of M and CO_2 (in water), then:

$$S_N - S_M = S_M - S_M + S_{CO2}\text{ (in water)} - S_{CO2}\text{ (in gas)} = 121.3-213.6 = -92.3\ \text{(J/K/mol)}$$
$$\Delta H = RT\ln P_{CO2} + T\,(S_N - S_M - S_{CO2}) = RT\ln(1.6) - 92.3T = (3.9-92.3)\,T = -88.4\,T\ \text{(J/mol)}$$

Taking 40°C (313K) as the CO_2 removal temperature, ΔH will be:

$$\Delta H = -88.4*313 = -27.67 \text{ (kJ/molCO}_2\text{)}$$

This is the typical ΔH required for an aqueous solvent based CO_2 removal process. For non-aqueous solvent this value may be even smaller because CO_2 will have higher entropy in the solution. It is interesting to note that this ΔH is close to the ΔH of aqueous carbonate/bicarbonate systems (27.2 kJ/molCO_2) [2], indicating that a carbonate/bicarbonate solution could be a promising solvent for CO_2 removal from syngas. Also a physical solvent or a mixed physical/chemical solvent with similar ΔH could be good candidates as well.

Carbonate/Bicarbonate Solutions

Several carbonate/bicarbonate systems, such as $K_2CO_3/KHCO_3$, $Na_2CO_3/NaHCO_3$, and $(NH_4)_2CO_3/(NH_4)HCO_3$ are available. These systems will have similar heat of absorption as long as there is no crystallization of any bicarbonate. Among these systems, ammonia carbonate ($(NH_4)_2CO_3$) is probably the most promising because it potentially has the highest CO_2 absorption capacity due to its low molecular weight of ammonia. $(NH_4)HCO_3$ also has the highest CO_2 equilibrium pressure at the CO_2 loading and when regenerated it will have the highest CO_2 decomposition pressure, which can reduce the compression work.

Ammonia carbonate is more challenging for post-combustion CO_2 removal (versus pre-combustion CO_2 removal—IGCC) due to the low partial pressure of CO_2 in the flue gas. When CO_2 partial pressure is low, especially at the top of the absorption column where the removal rate is high (say 90% capture), the pH of the carbonate/bicarbonate solution will be high, which will increase the ammonia vapor slip in the absorption column. In syngas the partial pressure of CO_2 is high (even at 90% removal the CO_2 partial pressure will be still at 1.6 atm) and the high CO_2 pressure will reduce the ammonia slip significantly.

Physical solvent based absorption processes

Current physical solvents-based absorption processes may be modified to improve their energy performance. For example, current Selexol process uses four staged flashed drums to release CO_2 and the average pressure of CO_2 is around 4 atms. However, one may use temperature swing regeneration to improve the energy performance. For thermal regeneration, most of the physical solvents may be too weak and a stronger physical solvent or a physical/chemical mixed solvent may improve the energy performance. It is possible that a well formulated mixed solvent will have better energy performance than the current Selexol process.

Conclusions

- Hot/warm CO_2 removal for syngas are energy intensive, a low temperature process maybe more attractive.
- Carbonate/bicarbonate solution or physical/chemical mixed solvents could be good candidates for CO_2 removal from syngas.

References

1. CRC Handbook of Chemistry and Physics, CRC Press, 60th edition, 1979.
2. J. T. Yeh, K. P. Resnik, K. Rygle, H. W. Pennline, *Fuel Processing Technology*, 86 (2005) 1533– 1546.

Novel Integration of Gas Separation Membranes for CO_2 Capture from IGCC Power Plants

Jared Ciferno, U.S. Department of Energy

John J. Marano, Ph.D., Consultant

National Energy Technology Laboratory, Pittsburgh, PA, USA

Abstract

A core mission of the U.S. Department of Energy's (DOE) Carbon Sequestration Program is to foster the development of commercially-ready technologies for CO_2 capture and sequestration. The DOE's Coal Fuels and Hydrogen Program has a similar mission for the production of hydrogen for use with advanced fuel-cells. R&D goals have been established for electric power generation from next-generation Integrated Gasification Combined Cycle (IGCC) plants and for future coal-to-hydrogen fuel plants. The missions of the two programs are intertwined in that both involve the production of hydrogen from coal using advanced coal gasification and synthesis gas clean-up systems.

Currently for carbon capture, the best available coal gasification system employs a two-stage physical-absorption system for H_2S removal and CO_2 capture. Coal is fed to a high-efficiency gasifier, and the conversion of CO in the raw syngas to H_2 is maximized in a water-gas shift (WGS) reactor train. For IGCC applications, the acid gas removal (AGR) system located downstream of the WGS produces a hydrogen-rich fuel gas that is combusted in a gas turbine (GT) topping cycle. H_2S removed in the AGR is sent to a Claus sulfur plant; while the CO_2 recovered is dehydrated and compressed for transport to an appropriate sequestration site. When hydrogen is the desired product, the optimal design is slightly different; less shifting of the raw syngas is required and the AGR is followed by a pressure swing adsorption (PSA) system to recover high-purity hydrogen for compression and sale. The by-product gas from the PSA, which contains H_2, CO and CO_2, is combusted with oxygen in a combined-cycle power plant to generate electric power for internal use. The CO_2 from the AGR and the exhaust from the combined-cycle is dehydrated and compressed for transport and sequestration.

R&D supported by the U.S. DOE is investigating alternatives to absorption for capturing CO_2 that may exceed the performance of existing technologies and achieve the DOE program goals. Membrane gas separation has been touted as one possible approach. Membranes have a number of advantages, in that they are usually compact, have no moving parts, have low maintenance, and can be highly reliable. The National Energy Technology Laboratory (NETL) is investigating alternative flowsheets incorporating membranes that may out-perform current technologies for CO_2 capture. An initial screening study has identified several novel integrations of membranes for IGCC and coal-to-hydrogen applications. For IGCC, the use of a N_2 sweep gas along with integration of the membrane within the CO_2 compression train is particularly attractive since high-purity hydrogen is not required for power generation. These options coupled with bulk H_2 recovery from the syngas after the WGS, can also produce a high-purity side stream for use in other applications. PSA can be can be used for bulk separation with H_2 recoveries in the range of 50% to 80%. This plant configuration allows for maximum flexibility in either the production of hydrogen for sale, or of power generation and sale, or both. Other potential advantages of this approach in addition to the elimination of the two-stage absorption unit are:

- The hydrogen membranes operate at a high feed pressure with a large partial-pressure differential across the membrane
- The use of a sweep gas improves the driving force across the membrane and also serves as the GT diluent
- Hydrogen re-compression is eliminated and the hydrogen fuel and sales gas are delivered at the required pipeline and turbine pressures, respectively
- The membrane is placed well downstream of the gasifier at a location where most of the contaminants in the raw syngas have been removed

A major challenge for implementing these CO_2 capture strategies is the development of membrane materials with high selectivities for H_2 relative to CO_2. The IGCC process with CO_2 capture using gas separation membranes, along with a similar process co-producing hydrogen and power, have been modeled with the Aspen process flowsheet simulator. Cases were developed for both warm and cold gas clean-up, and for no gas clean-up, where both CO_2 and H_2S are co-sequestered. Based on the results of this modeling, membrane development cost and performance targets are being developed.

Introduction

Figure 1 identifies possible integrations of gas separation membranes into an advanced IGCC power plant.

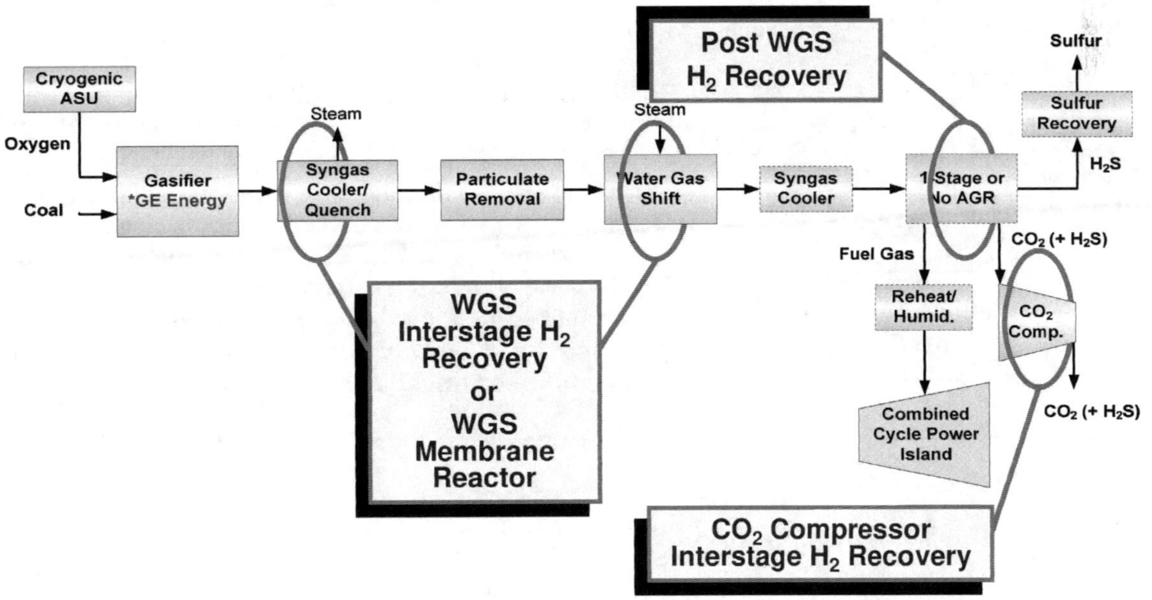

Figure 1. Integration of Membrane-Based Gas Separations with IGCC

Process conditions, gas composition, pressure and temperature, are different at the various locations identified in Figure 1. Intuitively, the operating envelope for any given membrane technology should match the conditions where it is being placed in the process. Also, each location in the flowsheet has its own unique advantages and disadvantages in regards to hydrogen separation and recovery. In addition, other technologies under development, such as warm-gas clean-up systems, may impact selection of a H_2/CO_2 separation technology, and may or may not complement membrane separation. Proper placement of a membrane unit in the process flowsheet is critical, and it is unlikely that a single membrane material can perform adequately at all feasible locations in the process. Therefore, the challenge is to take advantage of the unique characteristics of individual membrane technologies, while mitigating any shortcomings.

Analysis

The methodology used to integrate gas separation membranes into the IGCC process is given below:

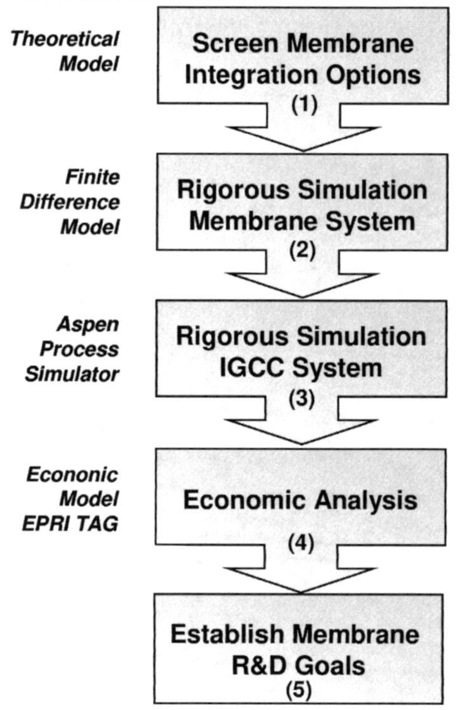

1. **Simple theoretical H_2 recovery model used to screen possible membrane locations and operating conditions**
2. **Candidate membrane configurations simulated using membrane unit model to determine membrane selectivity required to meet CO_2 capture target for different levels of H_2 recovery**
   - product of membrane permeance and area also determined
3. **Process system model used to simulate IGCC with membrane-based carbon capture; predicts:**
   - IGCC efficiency
   - equipment sizing
4. **Aspen results used in IGCC economic model to determine membrane cost required to meet Program target for COE**
5. **Establish quantitative application-specific R&D goals for membrane selectivity, permeance and cost**
   - selectivity target directly related to Program CO_2 capture goal
   - permeance and cost are directly related to Program COE goal

   Provide qualitative guidance on how to achieve R&D goals for selectivity, permeance and cost for specific membranes under development

The analyses of four different configurations have been completed:

- Configuration Ia – Membrane Integration Post-AGR with Staged H_2 Re-Compression
- Configuration Ib – Membrane integration Post-AGR with Permeate Sweep

- Configuration IIa – Membrane Integration with CO_2 Compression and Permeate Sweep
- Configuration IIb – Hybrid System with PSA & Membrane Integration with CO_2 Compression and Permeate Sweep

Process flow diagrams for these configurations are given at the end of the extended abstract. For each configuration, different cases were run based on the type of acid-gas (H_2S) removal employed: Cold-Gas AGR, Warm-Gas AGR, No AGR, (co-sequestration of CO_2 and H_2S).

Configuration Ia replaces conventional absorption technology with membranes. While less CO_2 compression is required, H_2 re-compression is necessary since the H_2 produced by the membrane unit is at low pressure. Re-compression is staged counter current to the membrane units to minimize the total amount of re-compression required. Performance of this system is comparable with conventional technology. Configurations Ib and IIa eliminate H_2 re-compression by strategic placement of multiple membrane units. In Ib, a sweep gas is used to enhance the recovery of H_2 without lowering the total pressure of the H_2-rich stream leaving the membrane system. Nitrogen from the air separation unit is used as the sweep gas, and also serves as the diluent to the gas turbine feed gas. Configuration IIa takes this concept one step further, by distributing the membrane units both post acid-gas removal and between CO_2 compression stages. The bulk of the H_2 recovery occurs before compression and incremental recovery is accomplished at the higher pressures available during the CO_2 compression for transport and sequestration.

Key performance metrics from the analysis of Configurations Ia, Ib, and IIa are reported in Table 1. For this analysis, the state-of-the-art (SOTA) absorption technology can achieve almost complete H_2 recovery; however, CO_2 recovery is at relatively low pressure. The membrane-based systems at best can only achieve about 98% H_2 recovery. A recovery of 95% is probably a practical upper limit for Ia and IIa. However, Configuration Ib is only able to achieve 90% H_2 recovery at the design conditions considered in the analysis.

Table 1. Comparison of Different Membrane Configurations with State-Of-The-Art CO_2 Capture Technology

	SOTA	**Configuration Ia**		**Configuration Ib**		**Configuration IIa**	
	CGCU	CGCU	WGCU	CGCU	WGCU	CGCU	WGCU
H_2 Recovery	99+%	95%	95%	90%	N/A	95%	N/A
H_2 Purity	91%	93%	61%	43%	N/A	44%	N/A
H_2 Recovery Pressure, psia	725	50-216	50-216	450	450	450	450
CO_2 Recovery	90%	90%	90%	90%	90%	90%	90%
CO_2 Purity	99+%	89%	87%	82%	N/A	88%	N/A
CO_2 Recovery Pressure, psia	50/150	745	745	745	745	745	745
Required H_2/CO_2 Selectivity	-	68	49	68	N/A	56	N/A
Relative Membrane Area	-	1.00	0.82	1.04	N/A	0.56	N/A

SOTA - State-Of-The-Art   CGCU - Cold-Gas Clean-Up
WGCU - Warm-Gas Clean-Up

All of the membrane configurations recover CO_2 at a much higher pressure than possible in the SOTA case. However as noted above, for Configuration Ia, H_2 recovery is achieved at lower pressures, off-setting this advantage. Configurations Ib and IIa are able to recover the H_2 at the desired pressure. Configuration IIa requires the smallest membrane area. Membrane selectivity requirements vary from 49 to 68, with lower selectivities (and smaller areas) required for systems incorporating warm-gas clean-up. This results from the condensation and removal of water in the cold-gas system. In the warm-gas system, this water is preferentially removed in the membrane units and further serves to dilute the H_2 permeate and improve the driving force across the membranes.

Configuration IIb is a hybrid system which combines membranes with pressure-swing adsorption (PSA). Each of these gas separation technologies has its own distinct advantages and disadvantages, and they can be used together to improve the performance of the overall gas separation system. In particular, Configuration IIb is advantageous in situations where the gasification plant co-produces high-purity H_2 for sale. The PSA unit performs the bulk of the hydrogen separation upstream of the membranes, producing high-purity hydrogen. Results of the analysis of Configuration IIb indicate that between 50 and 80% of the total hydrogen produced at the plant could be produced at purities exceeding 99%. While hydrogen re-compression is still not required for the fuel gas, use of PSA limits the operating pressure of the upstream gasifier and results in additional stages of CO_2 compression, depending upon the operating pressure selected for the PSA desorption cycle. Total H_2 recovery for this process configuration is similar to that achieved in Configuration IIa.

The results of this analysis are being used to generate performance and cost targets for DOE-sponsored membrane R&D. These targets, which relate directly back to the Carbon Sequestration Program goals, will be used by DOE project mangers and researchers to measure progress, and to provide guidance on future research directions for improving cost and performance. The development of these membrane R&D targets will be the subject of a future technical paper by the authors.

Conclusions

In conclusion, NETL analyses have shown that modular gas separation membranes may be attractively integrated into a number of locations in the IGCC process to separate CO_2 from H_2. More than one type of membrane material will likely be needed, if membranes are to be placed in multiple locations in the process. Syngas cooling, water-gas-shift conversion, syngas clean-up, and CO_2 capture for sequestration operate over a wide range of temperatures (40-1,000°C). The desired operating temperature for the membrane is a key factor in screening membrane materials for IGCC and hydrogen production applications. Low-temperature membrane applications should not be ruled out for CO_2 capture and may be more achievable in the short term.

High membrane selectivity is a prerequisite for successful process integration. H_2/CO_2 selectivity is directly related to the 90% CO_2 capture goal of the DOE R&D Program, and is still a major hurdle for most gas separation membranes under development. Based on this analysis, the minimum H_2/CO_2 selectivity appears to be on the order of 50. Other issues also remain to be resolved for membranes currently under development. Compared to commercial polymer membranes, cost and footprint will be issues with advanced membranes, and the cost of selective membrane materials and porous supports may be critical.

In summary, integrating gas separation membranes into the CO_2 compression train (Configuration IIa) has many benefits, since H_2 and recycle gas re-compression are undesirable for IGCC applications. Major advantages of this integration are the elimination of the 2nd-stage of AGR (and the 1st-stage if H_2S is co-sequestered), chemical solvents can be used for absorption of H_2S, CO_2 is delivered to compression at a much higher pressure, no H_2 re-compression is required, the H_2 separation membranes operate at very high pressures with large membrane pressure differentials, and the sweep gas further improves the pressure driving force and serves as a gas turbine fuel diluent. With further integration of a PSA unit with gas separation membranes, the coal gasification plant can also co-produce high-purity hydrogen for application in fuel cells. Between 50 and 80% of the hydrogen could be produced at high-purity in this hybrid gas-separation system (Configuration IIb).

Finally, NETL has begun analyses for other membrane/IGCC integrations. Figure 1, also identifies the option where membranes are integrated with the water-gas-shift reaction, up-stream of the acid gas removal system. This location has a number of advantages. Most notably, removal of hydrogen from the syngas before and during the water-gas-shift improves the conversion of CO and H_2O to H_2 and CO_2. Results from these analyses are anticipated by Fall of 2008.

Acknowledgements

The authors would like to express their appreciation to the U.S. DOE National Energy Technology Laboratory for providing financial support for this project.

Disclaimer

Any conclusions, comments or opinions expressed in this paper are solely those of the authors and do not represent any official position held by NETL, DOE, or the U.S. Government. Information contained herein has been based on the best data available to the authors at the time of preparation of the paper.

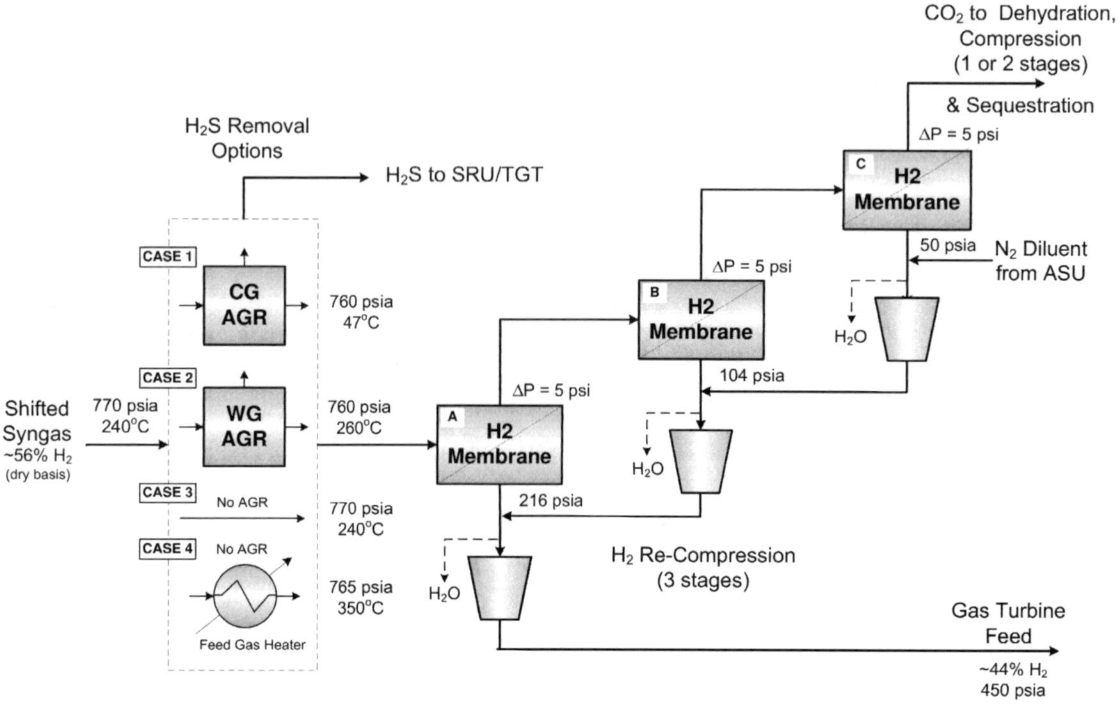

Figure 3a. Configuration Ia – Membrane Integration Post-AGR with Staged H_2 Re-Compression

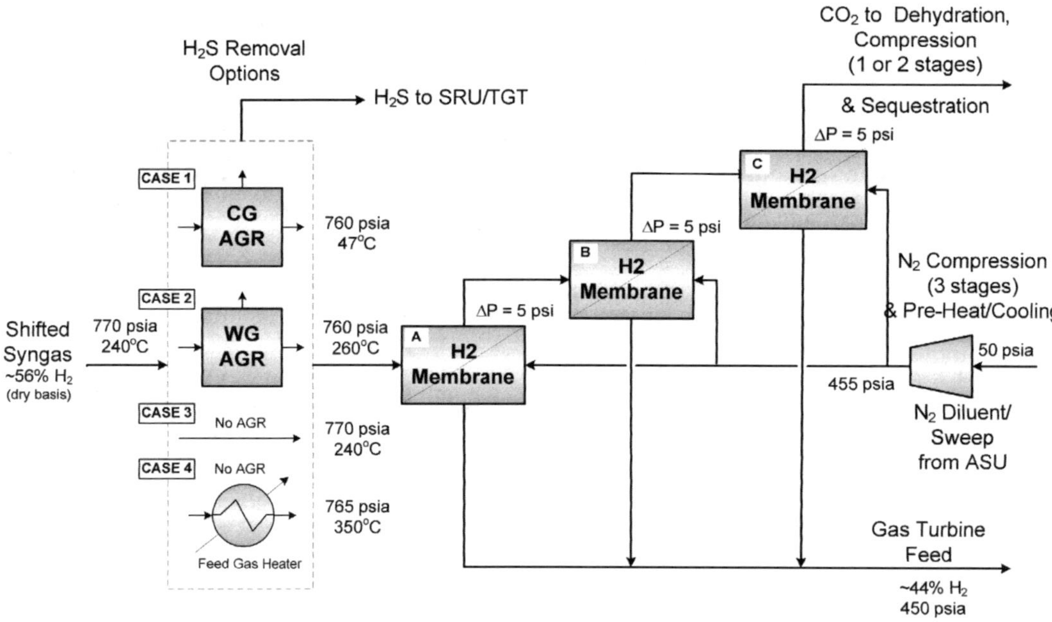

Figure 3b. Configuration Ib – Membrane Integration Post-AGR with Permeate Sweep

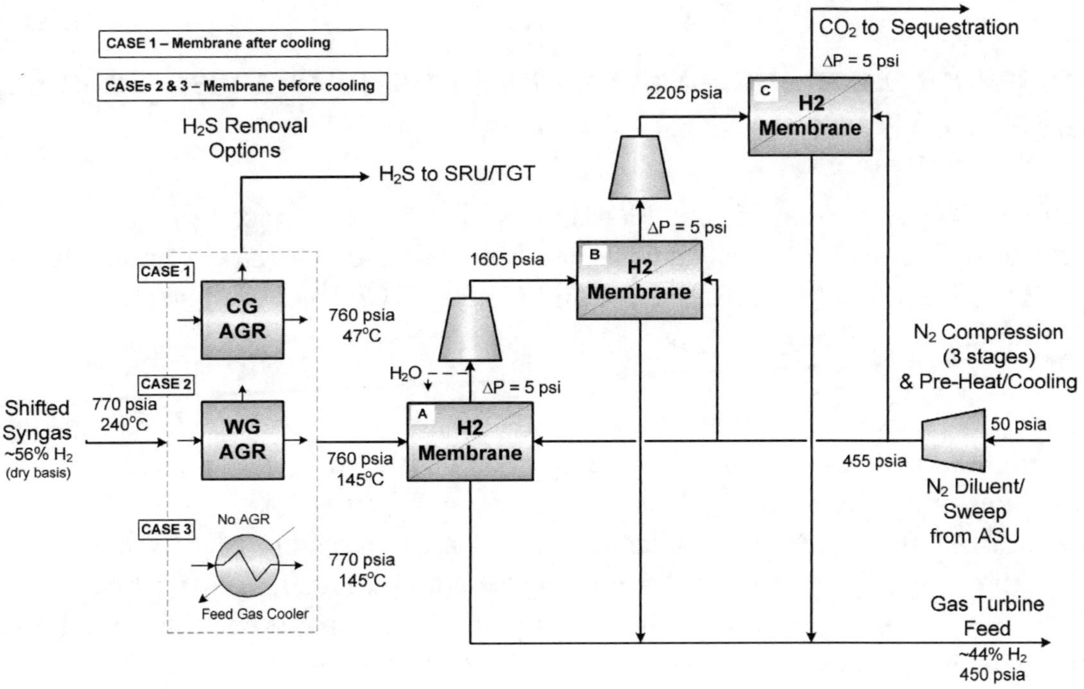

Figure 3c. Configuration IIa – Membrane Integration with CO_2 Compression and Permeate Sweep

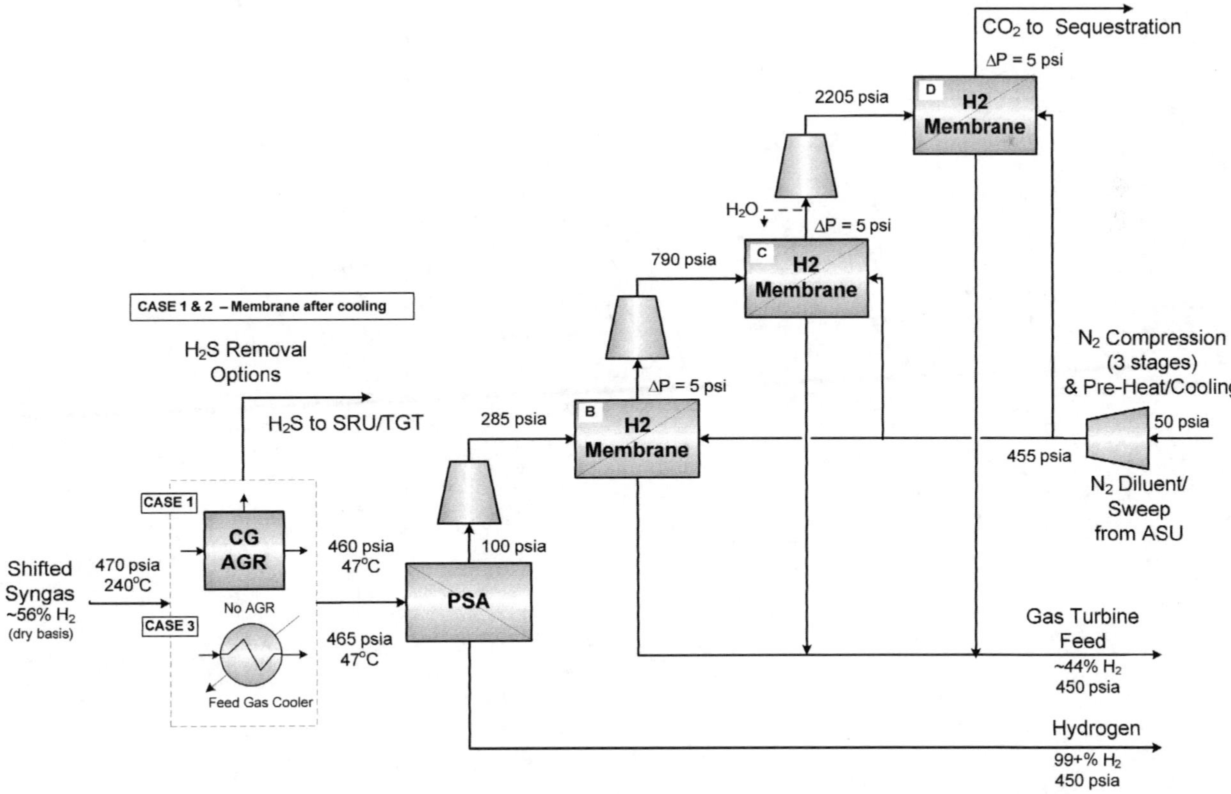

Figure 3d. Configuration IIb – Hybrid System with PSA & Membrane Integration with CO_2 Compression and Permeate Sweep

91g Technical and Economic Assessment of An Advanced Solid Oxide Fuel Cell-Based Power System

Eric Grol, Office of Systems Analysis and Planning, NETL, 626 Cochrans Mill Road, Pittsburgh, PA 15236, Phil Dipietro, Office of Sysems Analysis and Planning, NETL, 626 Cochrans Mill Road, Pittsburgh, PA 15236, and Dale Keairns, American Institute of Chemical Engineers, 3 Park Avenue, 19th Floor, New York, NY 10016.

Technical and economic evaluation of an atmospheric-pressure solid oxide fuel cell capable of achieving strict carbon capture targets, while maintaining high system efficiency, is presented. In this design syngas is passed once-through the fuel cell stack; residual fuel in the anode gas is combusted using a slip stream of oxygen from the cryogenic air separation unit. The high quality heat from the combusted anode gas is combined with heat recovered from syngas cooling to feed a supercritical steam bottoming cycle. The fuel cell stack produces 80% of the power with the remainder coming from the steam cycle. Key fuel cell assumptions include: 0.7 volts, 500 mA/cm2 current density, and 85% fuel utilization in the fuel cell. The proposed system design has the following advantages (1) it achieves 46% coal conversion efficiency (HHV) without recycle or excessive heat recovery, (2) all the system components save the fuel cell stack are commercially available today, (3) it exhaust a compression-ready stream of 95% pure CO_2, and (4) it allows capture and sequestration of nearly 100% of the carbon in the coal feedstock.

91h Decarbonisation in Energy Production

Yuhang Lou, Robin Smith, and Jhuma Sadhukhan. Centre for Process Integration, Chemical Engineering & Analytical Science, The University of Manchester, Manchester, United Kingdom

This work tackles the carbon dioxide (CO2) emission problem in relation to power generation.

Currently, CO2 emissions from power generation make up one of the largest contributions to global greenhouse gas emissions. For power generation, the CO2 emission is centralised, with a large volume, but low in concentration. There are three basic decarbonisation routes to deal with this problem. These are pre-combustion, post-combustion and the oxy-combustion routes. For each route, different CO2 separation technologies can be exploited. This work studies different decarbonisation routes with both conventional CO2 separation technologies and novel pre-combustion routes. Based on this, further studies on the issue of the CO2 recovery have also been carried out.

In conclusion, a wide range of decarbonisation designs can be applied with power generation. Among them, the oxy-combustion and the pre-combustion with novel CO2 separation technology have the best performances. The site power generation loss due to the CO2 capture is roughly proportional to the CO2 recovery. This indicates that to treat the CO2 recovery as a degree of freedom might be more sensible than as a fixed specification in the decarbonisation power generation design as is the case at present.

141a Mercury Control Strategy in Power Generation

Bruce A. Keiser, Engineered Materials Group, Nalco Company, 1601 West Diehl Road, Naperville, IL 60563 and Jitendra Shah, Separations Group, Nalco Company, 1601 West Diehl Road, Naperville, IL 60563.

Abstract:

The industrial revolution changed the way we think of manufacturing and made available to everyone a long list of items that make our lives easier. Many of these conveniences are powered by electricity. Today, the burning of coal generates about half of the electricity consumed in the US1. Estimates put the amount of coal reserves in North America at levels sufficient to provide electricity for the next 250 years.2 The burning of coal to produce electricity also produces solid waste, such as fly ash, and flue gas emissions into the atmosphere. Some estimate that two thirds of the SOx and a quarter of all the NOx in our atmosphere comes from electric power generation achieved by burning coal and other fuels.3 Besides the primary gas emissions, secondary emissions include such pollutants as arsenic, mercury, selenium, and boron. The Clean Air Act Amendments regulates many of these pollutants. Additionally, state legislations are taking a more stringent approach to the control of these pollutants. All of this creates a strong drive for the development and deployment of best available technology (BAT) to minimize the release of these byproducts of electricity generation. This presentation will focus on control strategies for mercury emissions in air and water process streams. Several technologies have been examined and tested over the last several years including activated carbon and inorganic sorbents. This paper will provide recent laboratory and field evaluations of additives that can be used to capture and prevent re-emission of mercury from electricity producing power plants.

References

1http://energy.careenergy.com/energy/energy-americas-economy.asp

2EIA, Annual Energy Review 2001, T.11.12, T.11.12, T.11.13 (http://www.careenergy.com/powering_life/abundant.asp)

3http://www.epa.gov/airmarkets/acidrain/index.html

Adsorbents for Reduction of Multiple Toxic Metals in Coal-based Power Generation

Henry G. Paris, Xing Dong, Steward Environmental Solutions, LLC , Glen Fryxell, Pacific Northwest National Laboratory

A major challenge for clean power using coal-based fuel is implementing techniques of use for adsorbents that can remove several toxic metals that exist in several valence states in the flue gas or the gas fuel. In smaller coal-fired plants that do not have an FGD it is advantageous to capture mercury using an adsorbent with high capacity for Hg^0 that works at ~150°C or higher and can be recovered and reused. Sorbent recovery and reuse provides significant cost advantage to utility operation as it reduces the amount of adsorbent required, and yields a mercury-free fly ash that can be used in secondary markets rather than being disposed in land fills. In larger and newer systems equipped with an FGD it can be advantageous to remove the mercury prior to the FGD to avoid re-emission of Hg^0. If the adsorbent shows activity towards other metals or speciation states such as Hg^{2+}, As^{3+} , Se^{3+} or copper it has obvious benefit to the utility power generation process. Toxic metals such as arsenic and selenium pose special problems as they exist in multiple valences. Although some adsorbents are claimed to be suitably active towards multiple valence states (As^{3+} and As^{5+}, for example, the design of an effective (cost and efficiency) method to remove multiple metals with the simplest and least mass of adsorbent may require a technique using more than one adsorbent. This presentation reviews several newly introduced adsorbents based on SAMMS® technology that can remove multiple toxic metals in various species of valence, particularly one that can be recovered and reused multiple times dramatically reducing the cost of removing mercury from flue gas. SAMMS® technology (self-assembled monolayers on mesoporous supports) is a unique way for the materials scientist to create high capacity adsorbents that can match or exceed the capacity and efficiency of synthetic zeolites and powdered activated carbon, and to design adsorbents with specific metal activity.

While powdered activated carbon is widely used, alternative sorbents can achieve improved and advantageous performance. For example, Figure 1 shows the performance a SAMMS® - based adsorbent in a magnetic support. When tested in a dry nitrogen stream at room temperature (in the form of a fixed bed configuration with 0.64 sec. dwell time) a reduction of greater than 99% was observed up to the end of the test at 500 minutes run time. Because the sorbent can be collected magnetically and reused, the unit cost drops significantly.

This result is particularly significant because the adsorption of elemental mercury did not require the use of additional additives to promote oxidation of mercury, nor did it require the presence of COS as is claimed by other investigators to be necessary for adsorption to magnetite.

The same adsorbent was tested in a side stream of a coal-fired power plant burning PRB coal. The contact time is 1 second and the gas temperature was 150°C. The mass injection rate was 4 pounds/mmacf. Table 1 shows the remaining experimental details including the gas chemistry. The adsorbent performance (Figure 2) is the same as the performance of Darco LH-Hg. The THFG product achieved the same performance in spite of the fact that it has a higher bulk density, lower number density, and lower surface area. The functionalized SAMMS adsorbent had a surface area of about 110 m^2/g, a particle size distribution whose d_{50} is between 50 and 88 micrometers and a bulk density of ~ 1.2 g/cc.

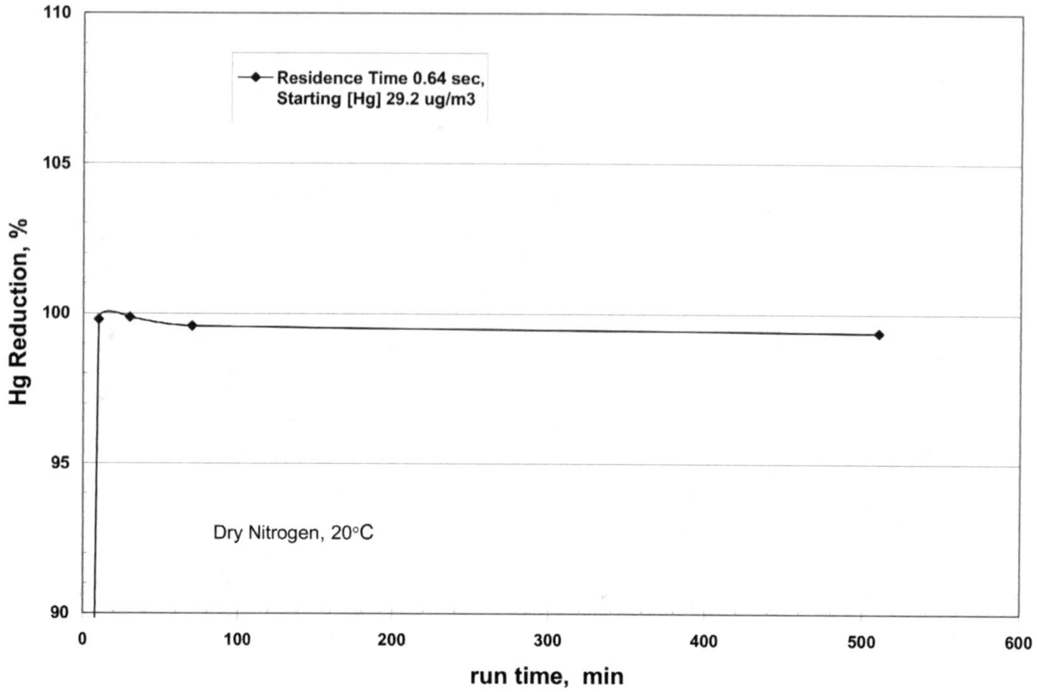

Figure 1. A laboratory simulation using dry nitrogen at room temperature achieved >99% reduction of mercury (GEM) for over 500 minutes.

Table 1. Boiler Test Conditions

	WKU Tests
Elemental Hg	7 μg/m^3
Other Hg	3 μg/m^3
Total Hg **Temperature**	10 μg/m^3 **151 °C**
Residence Time	1 sec
Gas Compositions	5~6% O_2; 15% COx; 100~300ppm SOx; 50~70ppm NOx; 7% H_2O; low chlorine

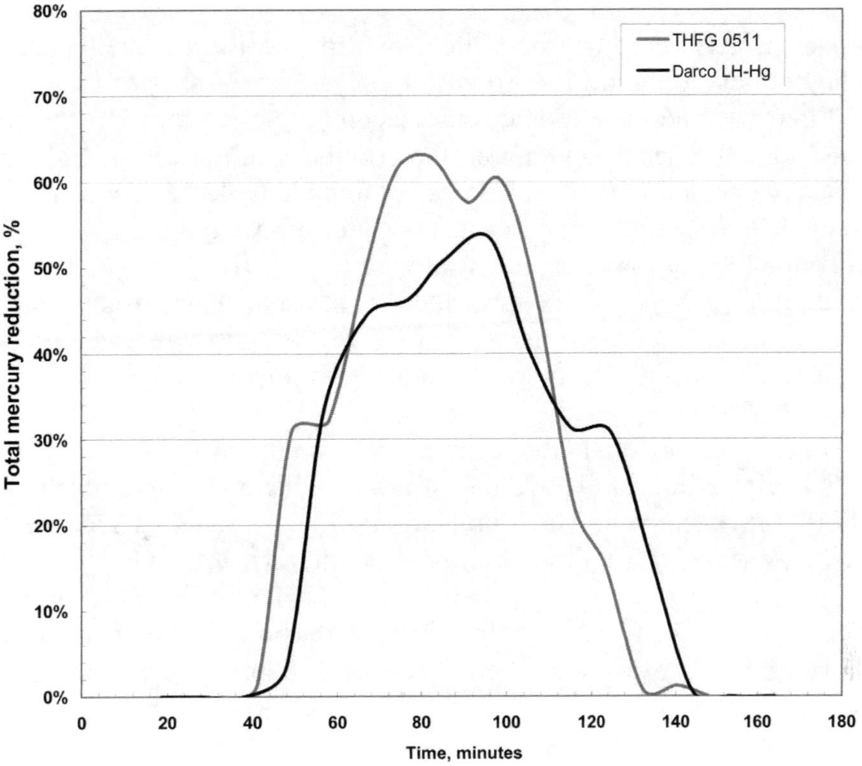

Figure 2. The total mercury reduction of Steward THFG and Darco LH-Hg tested in a PRB-fired utility boiler at 150°C, nominally 1 second dwell time. Injection began at 40 minutes.

PAC has a much smaller average particle size (95% <44 μm, d_{50} estimated at 25 μm) than the SAMMS adsorbent testing to Figure 2. The SAMMS adsorbent was further processed by screening. This screened adsorbent has a bulk density of ~1.2-1.4 g/cc. Table 2 shows the prepared material had a surface area of ~ 85-95 m^2/g and pore diameter of ~22 nm. Consider that typical PAC has a bulk density of ~ 0.6 g/cc, and has a surface area of ~ 500 m^2/g and both materials achieve the same reduction in total mercury. While there is significant difference in surface area relative to the starting material, there is little difference in surface area of the screened material. In a second test not shown here, the as-is achieved the same performance as the prepared material.

Table 2. Comparison of powder parameters for the THFG samples.

Sample	Surface area, m^2/g as-is	Surface area, -170m	Avg pore vol.c cm^3/g	D_{50} μm as-is	Avg. pore dia., nm	D_{50} μm -170m	D_{90}, μm as-is	D_{90}, μm -170m	Lab(RT) Hg reduction,% N_2
041707A-A	110	88	0.34	65	15.4	23	113	58	94
041707A-B	**110**	**94**	**0.37**	**36**	**15.5**	**24.5**	**44**	**58**	**93**
051507B	110	86	0.32	65	14.6	23	n/a	58	99
032907C	130	95	0.40	51	16.7	22	n/a	70	99
051107A	**130**	**95**	**0.36**	**51**	**14.8**	**21**	**n/a**	**52**	**99**

Since the adsorbent can be collected from the fly ash and reused multiple times, its drops the net cost to the utility proportionately to its number of reuse cycles. Based on the data in Figure 1, and an estimated one second exposure time in the flue, THFG could be used over 30,000 times depending on the efficiency of collection and extraction from the fly ash. Magnetic separation technology is readily available. It is used in the foodstuff industry to achieve very large removal percentages of ferrous particles.

Injection methods are inefficient in their use of adsorbent. With a residence time of ~ 1 second, the full capacity of the adsorbent is not used. Furthermore, mercury removal using injection methods can be strongly sensitive to the fluid dynamics of the gas flow and the properties of the adsorbent. A well thought out injection method is required to effectively utilize THFG SAMMS.

However, because THFG is recoverable from the process stream and a related magnetic SAMMS works in liquids (THFL-01), it is a candidate for use downstream of the ESP in the FGD liquid in a conventional application of adsorbents. Because the sorbent can be collected magnetically it will not compromise the gypsum. THFG SAMMS utilizes a thiol-ligand and has been shown in other work to bind selenite anions.

It is also possible to remove dissolved mercury from the effluent of the FGD using our silica-based thiol-SAMMS adsorbent THSL-01. In one study in the Great Lakes area using THSL-01 as a final polishing adsorbent, the FGD effluent was reduced to < 0.001 μg/L in six minutes contact time. The 2010 target of the Great Lakes Initiative is 0.0013 μg/L. Table 3 summarizes the results.

***Table 3**. Mercury Concentrations of THSL Treated FGD Effluent*

Contact time, minutes	[Hg] input, ppt	[Hg] output, ppt	Spiked, ppt	Duplicate, ppt	MS recdovery, %	RPD, %
0	4.499	-	8.266	7.348	98.6	11.8
6	-	0.687	2.381	2.529	93.4	6.0

Other SAMMS adsorbents are being developed based on anion SAMMS. These SAMMS® can remove oxometallate anions like chromate, arsenate and selenate to low levels, even in the presence of large excesses of sulfate ion. Nearly complete removal of arsenate and chromate has been achieved in the presence of interfering anions for solutions containing up to 100 ppm toxic metal anions under a variety of conditions. These anion-SAMMS® materials remain effective at even higher concentration solutions (in excess of 1000-ppm anions). The results of chromate adsorption support similar effect on arsenate and selenate.

Anion loading of more than 130 mg/g (1.12 mmol/g) of SAMMS® and distribution coefficients of more than 100,000 have been observed for chromate anion. These properties are comparable to the performance of thiol-SAMMS®. Anion removal tests were performed in water containing 1-, 10-, and 100-ppm arsenate and chromate with a water-to-sorbent (SAMMS®) ratio of 100. In each of these tests, virtually all of the chromate was removed in a single treatment. The presence of competing sulfate anions (at 150 ppm) had little effect on chromate binding. At the same solution-to-sorbent ratio (100 mL/g), chromate concentrations higher than 1000 ppm began to saturate the binding sites. For a much higher solution-to-silica ratio (500 mL/g), 100% removal of the chromate is observed for chromate concentrations up to 100 ppm.

In testing completed thus far, similar results were obtained for arsenate removal. The maximum loading capacity is 140 mg/g or 1.00 mmol/g. Under the same conditions, the residual concentrations of arsenate are all slightly higher than chromate at low anion concentrations. This suggests the Cu-EDA binding chemistry has somewhat higher affinity for chromate than arsenate under these conditions. An Fe-EDA SAMMS may be better.
The anion-SAMMS® materials are efficient oxometallate anion binding materials, virtually complete oxometallate anion removal can be achieved over a wide concentration range. The oxoanions (chromate and arsenate) are preferentially bound over sulfates and low concentrations of chromate and arsenate can be removed from high concentrations of sulfates. The binding chemistry is more sensitive for chromate than arsenate at low concentrations.

SAMMS adsorbent technology offers unique opportunities to achieve vey high reduction of mercury, arsenic and selenium in process streams from the coal-fired utility plant.

141c Palladium Sorbents for High Temperature Capture of Mercury, Arsenic, and Selenium from Fuel Gas

Evan J. Granite, United States Department of Energy, National Energy Technology Laboratory, PO Box 10940, M/S 58-106, Pittsburgh, PA 15236-0940

Abstract:

In gasification for power generation, the removal of mercury by sorbents at elevated temperatures preserves the high thermal efficiency of the integrated gasification combined cycle system. Unfortunately, most sorbents will display poor capacity for elemental mercury at elevated temperatures. Previous experience with sorbents in flue gas has allowed for judicious selection of potential high temperature candidate sorbents. The capacities of many sorbents for elemental mercury from nitrogen, as well as from different simulated fuel gases at temperatures from 400 -700 degrees F, were determined. The simulated fuel gas compositions contain varying concentrations of mercury, arsine, hydrogen selenide, carbon monoxide, hydrogen, carbon dioxide, moisture, and hydrogen sulfide.

Palladium is an attractive sorbent candidate for the removal of mercury from fuel gases at elevated temperatures. In addition, preliminary results suggest that palladium has excellent potential for arsenic and selenium capture from fuel gases, making it capable of multi-pollutant capture. A license agreement has been signed by the United States Department of Energy and Johnson Matthey for further development of the sorbents. Future sorbent development for trace metal capture from fuel gases will be discussed.

141d Uc Sulfur Recovery Process – An Integrated Multi-Contaminant Removal Process

Diana Matonis[1], Howard Meyer[1], Dennis Leppin[1], and Scott Lynn[2]. (1) Gas Technology Institute, 1700 S. Mt Prospect Rd, Des Plaines, IL 60018-1804, (2) University of California Berkeley, Berkeley, CA 94720

The University of California Sulfur Recovery Process (here after referred to as UCSRP-HP) testing addresses the development of an integrated multi-contaminant removal process in which H_2S, NH_3, HCl and heavy metals including Hg, As, Se and Cd present in a coal-derived syngas are to be removed to parts-per-million (ppm) or in some cases parts-per-billion (ppb) levels in a single process step. H_2S is converted directly into elemental sulfur at 275°F to 300°F and at the given sour gas pressure by reaction with sulfur dioxide in the liquid phase. Prior to this, the other contaminates such as NH_3, HCl and trace contaminants are removed in a separate section of the same reactor column. This is accomplished by the solvent absorbing H_2S, NH_3, HCl and trace metals from the feed gas. NH_3 and HCl form a highly soluble NH_4Cl salt and the absorbed heavy metals As, Cd and Hg precipitate out as their very insoluble sulfides. The Se trace metal, present in the syngas as H_2Se, forms a highly soluble $(NH_4)_2Se$ under these conditions and remains in the solvent. The proposed process is tightly integrated and is expected to be significantly more economical both in terms of capital and operating costs because it replaces the sulfur removal processes, acid-gas removal, Claus and SCOT, as well as the trace components used or proposed in conventional schemes by one single unit.

This paper will include an (i) investigation of long-term (i.e., 1000 hrs) solvent stability by exposure of the solvent to a simulated Illinois coal #6 gas containing all the contaminants that may be present in the feed to UCSRP reactor, (ii) investigation of metal corrosion related issues for selecting suitable material of construction for UCSRP reactor, (iii) investigation of the removal of the trace metal components from the syngas by the selected solvent, (iv) development of an Aspen-Plus based computer simulation model, and (v) techno-economic evaluation of the process applied to syngas cleanup.

AUTHOR INDEX

AUTHOR INDEX